McGraw-

500 College Biology Questions

Ace Your College Exams

Also in the McGraw-Hill Education 500 Questions Series

McGraw-Hill's 500 ACT English and Reading Questions to Know by Test Day
McGraw-Hill's 500 ACT Math Questions to Know by Test Day
McGraw-Hill's 500 ACT Science Questions to Know by Test Day
McGraw-Hill's 500 College Algebra and Trigonometry Questions: Ace Your College Exams
McGraw-Hill Education: 500 College Biology Questions: Ace Your College Exams
McGraw-Hill's 500 College Calculus Questions: Ace Your College Exams
McGraw-Hill's 500 College Chemistry Questions: Ace Your College Exams
McGraw-Hill's 500 College Physics Questions: Ace Your College Exams
McGraw-Hill's 500 Differential Equations Questions: Ace Your College Exams
McGraw-Hill's 500 European History Questions: Ace Your College Exams
McGraw-Hill's 500 French Questions: Ace Your College Exams
McGraw-Hill's 500 Linear Algebra Questions: Ace Your College Exams
McGraw-Hill's 500 Macroeconomics Questions: Ace Your College Exams
McGraw-Hill's 500 Microeconomics Questions: Ace Your College Exams
McGraw-Hill's 500 Organic Chemistry Questions: Ace Your College Exams
McGraw-Hill's 500 Philosophy Questions: Ace Your College Exams
McGraw-Hill's 500 Physical Chemistry Questions: Ace Your College Exams
McGraw-Hill's 500 Precalculus Questions: Ace Your College Exams
McGraw-Hill's 500 Psychology Questions: Ace Your College Exams
McGraw-Hill Education: 500 Regulation Questions for the CPA Exam
McGraw-Hill's 500 SAT Critical Reading Questions to Know by Test Day
McGraw-Hill's 500 SAT Math Questions to Know by Test Day
McGraw-Hill's 500 Spanish Questions: Ace Your College Exams
McGraw-Hill's 500 U.S. History Questions, Volume 1: Ace Your College Exams
McGraw-Hill's 500 U.S. History Questions, Volume 2: Ace Your College Exams
McGraw-Hill's 500 World History Questions, Volume 1: Ace Your College Exams
McGraw-Hill's 500 World History Questions, Volume 2: Ace Your College Exams

McGraw-Hill's 500 MCAT Biology Questions to Know by Test Day
McGraw-Hill's 500 MCAT General Chemistry Questions to Know by Test Day
McGraw-Hill's 500 MCAT Organic Chemistry Questions to Know by Test Day
McGraw-Hill's 500 MCAT Physics Questions to Know by Test Day

McGraw-Hill Education: 500 GMAT Math and Integrated Reasoning Questions to Know by Test Day
McGraw-Hill Education: 500 GMAT Verbal Questions to Know by Test Day
McGraw-Hill Education: 500 GRE Verbal Questions to Know by Test Day

McGraw-Hill Education

500 College Biology Questions

Ace Your College Exams

Robert Stewart

New York Chicago San Francisco Athens London Madrid
Mexico City Milan New Delhi Singapore Sydney Toronto

Copyright © 2015 by McGraw-Hill Education. All rights reserved. Printed in the United States of America. Except as permitted under the United States Copyright Act of 1976, no part of this publication may be reproduced or distributed in any form or by any means, or stored in a database or retrieval system, without the prior written permission of the publisher.

1 2 3 4 5 6 7 8 9 10   QFR/QFR   1 2 1 0 9 8 7 6 5

ISBN   978-0-07-178959-2
MHID   0-07-178959-6

McGraw-Hill Education products are available at special quantity discounts to use as premiums and sales promotions or for use in corporate training programs. To contact a representative, please visit the Contact Us pages at www.mhprofessional.com.

This book is printed on acid-free paper.

CONTENTS

Chapter 1   **Cell Chemistry**   1
Questions 1–21

Chapter 2   **Cell Structure and Function**   7
Questions 22–42

Chapter 3   **Cell Membrane and Function**   11
Questions 43–63

Chapter 4   **Enzymes**   15
Questions 64–84

Chapter 5   **Catabolic Metabolism**   19
Questions 85–105

Chapter 6   **Photosynthesis**   25
Questions 106–126

Chapter 7   **Mitosis**   29
Questions 127–147

Chapter 8   **Meiosis**   35
Questions 148–168

Chapter 9   **Inheritance**   39
Questions 169–189

Chapter 10   **DNA Replication and Repair**   43
Questions 190–210

Chapter 11   **RNA Transcription**   47
Questions 211–231

Chapter 12   **Translation**   53
Questions 232–252

Chapter 13   **Genetic Engineering**   59
Questions 253–273

Chapter 14   **Origin of Life**   65
Questions 274–294

Chapter 15   **Evolutionary Mechanisms and Speciation**   71
Questions 295–315

Chapter 16 **Viruses** 77
Questions 316–336

Chapter 17 **Prokaryotes** 81
Questions 337–357

Chapter 18 **Protozoans** 85
Questions 358–378

Chapter 19 **Fungi** 91
Questions 379–399

Chapter 20 **Plants** 97
Questions 400–420

Chapter 21 **Animals** 101
Questions 421–440

Chapter 22 **Ecological Principles** 105
Questions 441–460

Chapter 23 **Population Ecology** 109
Questions 461–480

Chapter 24 **Communities and Ecosystems** 113
Questions 481–500

Answers 119

CHAPTER 1

Cell Chemistry

1. The monomers that are used to build the polymers we identify as proteins are
   (A) fatty acids.
   (B) amino acids.
   (C) monosaccharides.
   (D) vitamins.
   (E) nucleotides.

2. Cellulose is a polysaccharide composed of many molecules of glucose. Yet most animals can ingest cellulose and never harvest a single calorie of energy for their cellular metabolism. Why is this?
   (A) Most animals are incapable of eating the trees in which the cellulose is found.
   (B) Animals can ingest cellulose but only in very limited amounts.
   (C) The form of glucose used to make cellulose is different than the form of glucose normally acquired by animals.
   (D) Most animals lack the enzyme needed to break apart cellulose.
   (E) The glucose is inaccessible because cellulose is a highly branched molecule.

3. If adding 10 drops of an acid into a solution changes the pH of the solution from 5.0 to 4.0, what would happen if an additional 10 drops were added?
   (A) The pH of the solution would return to 5.0.
   (B) The pH of the solution would drop to 3.0.
   (C) The pH of the solution would drop to approximately 3.9.
   (D) It is impossible to predict the final pH of the solution.
   (E) The pH of the solution would remain unchanged because it would still be equilibrating from the first 10-drop adjustment.

4. Which of the following is NOT a characteristic of proteins?
   (A) They are associated with regulating cellular activities.
   (B) They are capable of storing energy that can be used by a cell.
   (C) They are capable of participating in cellular movement.
   (D) They are commonly required to transport substances across a cell membrane.
   (E) They are associated with the storage of genetic information in cells.

5. Which of the following is NOT a good descriptive term of the structure of DNA?
   (A) parallel
   (B) base paired
   (C) double stranded
   (D) organic
   (E) semiconservatively replicated

6. Sulfur is an essential nutrient because it is required for the synthesis of
   (A) lipids.
   (B) proteins.
   (C) carbohydrates.
   (D) nucleic acids.
   (E) cell walls.

7. The cohesive ability of water is due to what molecular characteristic?
   (A) its small size
   (B) the ratio of hydrogen to oxygen
   (C) its low mass
   (D) its partial polarity
   (E) its low freezing point

8. *Amphipathic* is a term that would be well used to describe
   (A) a protein.
   (B) a nucleic acid.
   (C) a molecule of water.
   (D) an enzyme and cofactor interaction.
   (E) a phospholipid.

9. The four disulfide crosslinked chain structure of an antibody molecule is described as characterizing its _____ structure.
   (A) alpha
   (B) primary
   (C) quaternary
   (D) polarized
   (E) tertiary

10. Which of the following substances is most alkaline?
    (A) oven cleaner
    (B) vinegar
    (C) household bleach
    (D) blood
    (E) carbonated soft drinks

11. Which of the following are all hexoses?
    (A) glucose, mannose, sucrose
    (B) ribose, deoxyribose, glucose
    (C) dextrose, glucose, sucrose
    (D) mannose, galactose, dextrose
    (E) lactose, ribose, glucose

12. Which of the following is a hydrophobic molecule?
    (A) NaCl
    (B) CO_2
    (C) fatty acid
    (D) protein
    (E) sucrose

13. Why are halogens effective as disinfectants?
    (A) They tend to donate electrons to adjacent molecules that then distort their structures.
    (B) They remove oxygen from the environment, and organisms cannot survive.
    (C) They react strongly to any sodium in a solution and precipitate salt.
    (D) They are small, and cells cannot prevent them from penetrating the cell membranes.
    (E) They are strongly oxidizing and denature proteins.

14. What is the relationship between amylose, glycogen, and starch?
    (A) Amylose is a monosaccharide, while the other two are polysaccharides.
    (B) Glycogen, starch, and amylose are polymers of glucose.
    (C) Amylose and glycogen are found in animals, while starch is in plants.
    (D) Glycogen is composed of amylose, while starch is unrelated to the other two.
    (E) Glucose is broken down in order to form both glycogen and starch.

15. Buffers are chemicals with the ability to maintain a fairly constant pH in a solution even while acids or bases are being added. How does a buffer work?
    (A) Buffers absorb equal amounts of hydrogen and hydroxyl ions.
    (B) Buffers combine hydrogen and hydroxyl ions to form neutral water.
    (C) Buffers convert the two respective ions into gases that remain dissolved in the solution.
    (D) Buffers form a salt with the excess ions, which precipitates out of the solution.
    (E) Buffers convert the two respective ions into gases that escape the solution into the surrounding atmosphere.

16. What do the following have in common: bile salts, reproductive hormones, and cholesterol?
    (A) All three are steroids.
    (B) All three are manufactured from fatty acids.
    (C) All three have a role in regulating metabolism.
    (D) All three are strongly associated with the liver.
    (E) All three have a role in reproduction.

17. How are the cell walls of bacteria related to the cell walls of fungi and the exoskeleton of insects?
    (A) Bacterial cell walls are used as precursors to construct both the fungal cell wall and the exoskeletons.
    (B) All three are derived from a common amino acid sequence precursor.
    (C) All three are polysaccharides.
    (D) All three share a common core lipid-based structure.
    (E) The cell walls of fungi and the exoskeleton of insects share the same peptide crosslinking pattern seen in the bacterial cell wall.

18. What structure is common to all lipids?
    (A) having a single amino acid core
    (B) having a three-carbon glycerol head
    (C) having glycoside bonds linking fatty acids
    (D) having three fatty acids in each lipid molecule
    (E) containing at least one nitrogen or phosphorus atom

19. Which of the following is NOT characteristic of an amino acid?
    (A) has a central carbon atom
    (B) has an organic or hydrogen side group
    (C) contains nitrogen
    (D) has a carboxyl side group
    (E) is a ketone

20. Iodine is a halogen that is commonly used as a disinfectant. Yet it is also an essential nutrient. How are these two facts reconcilable?
    (A) Iodine is essential because it serves to protect cells from bacterial damage.
    (B) Iodine acts as a disinfectant when at low levels, but at higher levels it functions to regulate cellular mechanisms.
    (C) Iodine is capable of entering into and damaging bacteria but cannot pass through the cell membranes of higher organisms.
    (D) Iodine retards bacterial growth when at high levels but is required at low levels for the production of hormones that regulate metabolism in animals.
    (E) The presence of iodine in certain white blood cells enhances their ability to clear bacteria from the blood and tissues in animals.

21. If a carbon atom always forms four bonds, how can a molecule such as C_2H_4 exist?
    (A) The carbon atoms have phantom bonds that do not link to atoms when they are too far away but can if they get closer.
    (B) Carbon atoms do not necessarily have to have four bonds; sometimes they can rapidly alternate between three and five.
    (C) A double bond forms between the two carbon atoms.
    (D) This molecule cannot actually exist; it is hypothetical only.
    (E) The hydrogen atoms can bond to each other in this structure, making the required number of total bonds come out correctly.

CHAPTER 2

Cell Structure and Function

22. Which of the following best differentiates a fungal cell from a bacterial cell?
    (A) the presence of a cell wall
    (B) the proteins present in the cell membrane
    (C) the ability to metabolize toxic materials
    (D) the presence of a nucleus
    (E) the presence of photosynthetic pigments

23. The term *trilaminar* best refers to what cellular structure?
    (A) a cell wall
    (B) a cell membrane
    (C) the arrangement of chromosomes
    (D) the arrangement of microtubules within a flagellum
    (E) the composition of a ribosome

24. What do insects and fungi have in common?
    (A) Both are dimorphic.
    (B) Both are capable of anaerobic respiration.
    (C) Both have a substance called chitin.
    (D) Both produce the hormone estrogen.
    (E) Both forms require a soil-bound life stage.

25. Materials pass in and out of a nucleus by a process called
    (A) gated transport.
    (B) vesicular transport.
    (C) osmosis.
    (D) antiport.
    (E) symport.

26. The endoplasmic reticulum structure is continuous with the
    (A) Golgi apparatus.
    (B) cell membrane.
    (C) mitochondria.
    (D) nucleus.
    (E) ribosomes.

27. Which cellular structure is present in all cells?
    (A) a cell wall
    (B) a nucleus
    (C) mitochondria
    (D) a cytoskeleton
    (E) ribosomes

28. If the endoplasmic reticulum (ER) is analogous to a factory assembly line, what would be the analogy for the Golgi apparatus?
    (A) trucks
    (B) library
    (C) head office
    (D) loading dock
    (E) generator

29. What do plant cells have that no other cells have?
    (A) cell walls of chitin
    (B) mitochondria
    (C) plastids
    (D) trilaminar cell membranes
    (E) chromatin

30. What cellular structures provide protection from toxic materials?
    (A) peroxisomes
    (B) chromosomes
    (C) endosomes
    (D) lysosomes
    (E) ribosomes

31. Which of the following correctly arranges the items from the smallest to the largest in size?
    (A) virus – protein – ovum – mitochondria – nucleus
    (B) protein – virus – mitochondria – nucleus – ovum
    (C) protein – mitochondria – virus – nucleus – ovum
    (D) virus – protein – mitochondria – ovum – nucleus
    (E) nucleus – virus – protein – mitochondria – ovum

32. The ability of amoebas to move by extensions of pseudopodia is controlled by which of the following?
    (A) microtubules
    (B) sarcolemma
    (C) microfilaments
    (D) myofibrils
    (E) intermediate filaments

33. How are mitochondria and chloroplasts similar?
    (A) Both are about the same size as viruses.
    (B) Both have their own ribosomes.
    (C) Both are connected to the endoplasmic reticulum.
    (D) Both function in the Calvin-Benson cycle.
    (E) Both are associated with catabolic activities.

34. What should the term "80S" immediately bring to mind to the biologist?
    (A) the eukaryotic flagellum
    (B) the electron transport system
    (C) the membrane transport proteins for glucose
    (D) the intact eukaryotic ribosome
    (E) the nucleosome

35. What should the term "9+2" immediately bring to mind to the biologist?
    (A) the bacterial pilus
    (B) the cell membrane structure
    (C) the nucleus
    (D) chloroplasts and mitochondria
    (E) the eukaryotic flagellum

36. The ribosome is best associated with what process?
    (A) anaerobic respiration
    (B) catabolic mechanisms
    (C) translation
    (D) translocation
    (E) mitosis

37. What is first required for the assembly of the DNA polymerase holoenzyme on genomic DNA prior to DNA replication?
    (A) *Taq*
    (B) the α-subunit of DNA polymerase
    (C) the σ-subunit
    (D) the β-subunit of DNA polymerase
    (E) topoisomerase

38. If the DNA were extracted from a single human nucleus and the chromatin laid end-to-end, how long would the resulting molecule be?
    (A) about 300 mm
    (B) about 1.8 m
    (C) about 6.5 µm
    (D) about 10 cm
    (E) about 45 nm

39. Before mRNA can leave the nucleus for translation into proteins, what must first be accomplished?
    (A) Introns must be removed and the remaining exons spliced together.
    (B) A poly-A tail must be added at the 3' end of the message.
    (C) A 7-meG cap must be added to the 3' end of the molecule.
    (D) A, B, and C
    (E) both A and B

40. Which of the following is an example of an inducible operon?
    (A) *lac*
    (B) *trp*
    (C) *Eco*RI
    (D) *glu*
    (E) *Taq*

41. A proto-oncogene is
    (A) any DNA sequence that contains a recognition element.
    (B) any gene that causes cancer.
    (C) any normal essential growth regulating gene.
    (D) any sequence associated with mutations.
    (E) another name for a mobile genetic element.

42. DNA can be found in
    (A) a cell nucleus.
    (B) a ribosome.
    (C) a mitochondrion.
    (D) A, B, and C
    (E) both A and C

CHAPTER 3

Cell Membrane and Function

43. Phospholipids, when placed in water, may form clusters called
    (A) phospholipids.
    (B) membranes.
    (C) nonpolar clusters.
    (D) micelles.
    (E) envelopes.

44. Which of the following does NOT represent a form of active transport?
    (A) endocytosis
    (B) sodium-potassium pump
    (C) hydrogen ion transport with the gradient
    (D) exocytosis
    (E) hydrogen ion transport against a gradient

45. Classes of lipids in a cell membrane include
    (A) glycolipids.
    (B) cholesterol.
    (C) phospholipids.
    (D) all of the above
    (E) both A and C

46. Water passes through a membrane by
    (A) simple diffusion.
    (B) channel-mediated passive transport.
    (C) carrier-mediated passive transport.
    (D) carrier-mediated active transport.
    (E) none of the above

47. Components of a cell's endomembrane system include
    (A) endosomes and lysosomes.
    (B) mitochondria and chloroplasts.
    (C) the nucleus and the endoplasmic reticulum.
    (D) all of the above
    (E) both A and C

48. Beta-barrels are classified as _____ proteins.
    (A) lipid-linked
    (B) protein attached
    (C) transmembrane
    (D) none of the above
    (E) both A and C

49. Phospholipids vary
    (A) in the content of their heads.
    (B) in the length of their tails.
    (C) in the content of their tails.
    (D) all of the above
    (E) both B and C

50. Of the following, which will freely flow from higher to lower concentrations across a membrane?
    (A) O_2
    (B) H^+
    (C) proteins
    (D) nucleotides
    (E) both C and D

51. Proteins _____ move by vesicular transport.
    (A) moving from the cytosol to the nucleus
    (B) moving from the cytosol to mitochondria
    (C) moving from the endoplasmic reticulum to the Golgi apparatus
    (D) moving from the Golgi apparatus to secretory vesicles
    (E) both C and D

52. The flow of ions across a membrane is best associated with
    (A) α-helices.
    (B) peripheral membrane proteins.
    (C) β-barrels.
    (D) glycosylated proteins.
    (E) none of the above

53. Common motion(s) expressed by phospholipids in a bilayer include
    (A) lateral diffusion.
    (B) flip-flop.
    (C) flexion.
    (D) all of the above
    (E) both A and C

54. Sodium ions, with a gradient, can cross a membrane by
    (A) simple diffusion.
    (B) channel-mediated passive transport.
    (C) carrier-mediated passive transport.
    (D) carrier-mediated active transport.
    (E) none of the above

55. Disulfide bridges in membrane-associated proteins
    (A) will form only on the cytosolic leaflet.
    (B) will form only on the exterior leaflet.
    (C) will crosslink the protein to oligosaccharides on the cytosolic side.
    (D) will crosslink the protein to oligosaccharides on the exterior leaflet.
    (E) both B and D

56. Antiport is a form of
    (A) uniport.
    (B) symport.
    (C) coupled transport.
    (D) simple diffusion.
    (E) none of the above

57. The cell cortex is associated with
    (A) maintaining vesicle structure.
    (B) the exterior leaflet of the cell membrane.
    (C) maintaining endoplasmic reticulum structure.
    (D) the cytosolic leaflet of the cell membrane.
    (E) both A and C

58. Stabile membrane domains include
    (A) replicative rafts.
    (B) protein-attached glycolipids.
    (C) intracellular cortex aggregates.
    (D) tight junction bands.
    (E) none of the above

59. ATP driven pumps in animals are most commonly used to pump
    (A) H^+.
    (B) Na^+.
    (C) O_2.
    (D) Mn^{2+}.
    (E) none of the above

60. Glucose, against a gradient, can cross a membrane by
    (A) simple diffusion.
    (B) channel-mediated passive transport.
    (C) carrier-mediated passive transport.
    (D) carrier-mediated active transport.
    (E) none of the above

61. The hydrophobic tails of membrane phospholipids are most commonly _____ carbons in length.
    (A) 13–17
    (B) 18–20
    (C) 14–19
    (D) 19–24
    (E) 20–24

62. Protein import into a chloroplast proceeds in the following sequence:
    (A) signal sequence binds to receptor, protein-receptor complex moves laterally, protein refolds, signal sequence is removed.
    (B) signal sequence binds to receptor, protein refolds, protein-receptor complex moves laterally, signal sequence is removed.
    (C) signal sequence binds to receptor, signal sequence is removed, protein-receptor complex moves laterally, protein refolds.
    (D) signal sequence binds to receptor, protein-receptor complex moves laterally, signal sequence is removed, protein refolds.
    (E) none of the above

63. Signal transduction down a neuron's axon is conducted by _____ ion channels.
    (A) voltage gated
    (B) ligand gated
    (C) mechanically gated
    (D) both A and B
    (E) both B and C

CHAPTER 4

Enzymes

64. What enzyme is responsible for the production of tRNA?
    - (A) reverse transcriptase
    - (B) DNA polymerase
    - (C) DNA gyrase
    - (D) RNA polymerase
    - (E) restriction endonuclease

65. Which of the following does NOT affect enzyme function?
    - (A) pH
    - (B) temperature
    - (C) nitrogen
    - (D) cofactors
    - (E) inhibitors

66. An enzyme is composed of _____ subunits.
    - (A) fatty acid
    - (B) amino acid
    - (C) lipid
    - (D) simple sugar
    - (E) organic ion

67. The location on an enzyme where the actual catalytic activity resides is known as the
    - (A) prosthetic site.
    - (B) dehydration site.
    - (C) operon.
    - (D) active site.
    - (E) allosteric site.

68. The tertiary functional shape of an enzyme is maintained by the presence of _____ and _____.
    (A) hydrogen bonds; covalent bonds
    (B) covalent bonds; ionic bonds
    (C) hydrogen bonds; disulfide bridges
    (D) ionic bonds; disulfide bridges
    (E) covalent bonds; disulfide bridges

69. Beside proteins, what other biologic compound can possess some form of catalytic ability?
    (A) RNA
    (B) DNA
    (C) lipids
    (D) triglycerides
    (E) polysaccharides

70. In which of the following biologic functions is an enzyme NOT involved?
    (A) conversion of a single substrate molecule to two product molecules
    (B) splicing of two mRNA exons
    (C) conversion of two substrate molecules to one product molecule
    (D) production of DNA
    (E) binding of one cell to another

71. What is the primary function of an enzyme?
    (A) to release energy from a molecule
    (B) to release the energy stored within an ATP molecule
    (C) to store energy with a newly formed chemical bond
    (D) to convert chemical energy into biologic energy
    (E) to lower the activation energy required to form or break a chemical bond

72. Which of the following is NOT considered an enzyme?
    (A) DNA ligase
    (B) DNA-dependent RNA polymerase
    (C) acetyl CoA
    (D) transacetylase
    (E) reverse transcriptase

73. Which of the following molecular weights would be least likely to be linked to an enzyme's mass?
    (A) 600 kDa
    (B) 20 kDa
    (C) 128 kDa
    (D) 200 Da
    (E) 16,000 Da

74. Which of the following terms best describes the shape of an enzyme's active site?
    (A) a balloon
    (B) a cave
    (C) a bulge
    (D) a tab
    (E) a cave

75. Exergonic reactions are most closely linked to _____ processes.
    (A) synthetic
    (B) energy storing
    (C) catabolic
    (D) linking
    (E) genetic

76. Enzymes are
    (A) biologic catalysts.
    (B) organic compounds.
    (C) large globular proteins.
    (D) all of the above
    (E) both A and C

77. A cofactor is
    (A) a small enzyme.
    (B) an organic molecule needed to help an enzyme to function.
    (C) an ion that inhibits an enzyme.
    (D) an inorganic molecule that binds at an enzyme's allosteric site.
    (E) a molecule that causes feedback inhibition.

78. ΔG is a figure used to represent
    (A) a type of protein structure.
    (B) molecular bond angles.
    (C) the number of carbon atoms in a lipid.
    (D) differences in molecular energy levels.
    (E) none of the above

79. An inhibitor that binds at the active site of an enzyme is called
    (A) a competitive inhibitor.
    (B) a cofactor.
    (C) a noncompetitive inhibitor.
    (D) an allosteric inhibitor.
    (E) none of the above

80. If a cofactor is not present, then the activity of the specific enzyme will be
    (A) slightly increased.
    (B) greatly increased.
    (C) unaffected.
    (D) marginally reduced.
    (E) reduced to zero.

81. Which of the following is NOT an exergonic reaction?
    (A) muscle contraction
    (B) fire
    (C) digestion
    (D) lipid synthesis
    (E) respiration

82. The term *substrate* is most similar to
    (A) product.
    (B) cofactor.
    (C) reactant.
    (D) inhibitor.
    (E) enzyme.

83. In order for an enzyme to function, what must be added to the enzyme–substrate mix?
    (A) product
    (B) coenzymes
    (C) inhibitors
    (D) oxygen
    (E) energy

84. When a molecule is oxidized, what is actually happening?
    (A) A molecule is accepting an oxygen atom.
    (B) A molecule is losing an electron.
    (C) A molecule is losing a hydrogen ion.
    (D) all of the above
    (E) none of the above

CHAPTER 5

Catabolic Metabolism

85. Which is a five-carbon compound that is a portion of the TCA cycle?
    (A) succinyl CoA
    (B) acetyl CoA
    (C) oxaloacetate
    (D) pyruvate
    (E) α-ketoglutarate

86. The production of two molecules of pyruvate from a molecule of glucose is classified as a(n)
    (A) catabolic process.
    (B) synthetic reaction.
    (C) aerobic respiration.
    (D) anabolic process.
    (E) β-oxidation reaction.

87. Which of the following classes of compounds contain the most energy per gram?
    (A) fats
    (B) proteins
    (C) carbohydrates
    (D) monosaccharides
    (E) B, C, and D

88. Which of the following molecules contains the most available energy?
    (A) ATP
    (B) cAMP
    (C) glucose
    (D) ADP
    (E) NADH

89. Pyruvate can be regarded as the end product of
    (A) photolysis.
    (B) acetyl CoA conversion.
    (C) electron transport.
    (D) glycolysis.
    (E) the Krebs cycle.

90. An organism that uses a compound such as a protein for both its energy and carbon source would be classified as a(n)
    (A) photoautotroph.
    (B) chemoheterotroph.
    (C) photoheterotroph.
    (D) chemoautotroph.
    (E) autotroph.

91. _____ is a component of the mitochondrial electron transport chain.
    (A) Flavoprotein
    (B) Pyruvate
    (C) Cytochrome
    (D) both A and B
    (E) both A and C

92. What is the final electron acceptor in the mitochondrial electron transport chain?
    (A) ADP
    (B) ATP
    (C) CoQ
    (D) O_2
    (E) NAD

93. *Aerobic* refers to what property of chemotrophs?
    (A) Chemotrophs remove CO_2 from the air.
    (B) Chemotrophs require oxygen to produce energy.
    (C) Chemotrophs use but do NOT require oxygen.
    (D) Fermentation releases great amounts of gas, but only in chemotrophs.
    (E) Heat losses occur in the presence of oxygen.

94. Which of the following is a four-carbon molecule?
    (A) α-ketoglutarate
    (B) oxaloacetate
    (C) citrate
    (D) succinyl CoA
    (E) both B and D

95. The breakdown of glucose to pyruvate by a cell is an example of a(n) _____ reaction.
    (A) anabolic
    (B) aerobic
    (C) catabolic
    (D) synthesis
    (E) none of the above

96. In eukaryotes, aerobic respiration generates _____ ATP molecules from a molecule of glucose.
    (A) 2
    (B) 3
    (C) 36
    (D) 38
    (E) 11

97. Photoheterotrophs are best described as organisms that obtain energy to make ATP
    (A) from organic compounds but use sunlight to produce carbon sources.
    (B) from sunlight but cannot make organic compounds from CO_2.
    (C) and organic compounds from sunlight.
    (D) from some forms of chemicals.
    (E) from organic compounds.

98. The sequential process of fatty acid catabolism to acetyl CoA is called
    (A) hydrogenation.
    (B) oxidative phosphorylation.
    (C) the glyoxylate cycle.
    (D) reduction.
    (E) β-oxidation.

99. When NADH is converted to NAD, the process is categorized as
    (A) dehydration.
    (B) oxidation.
    (C) catalysis.
    (D) reduction.
    (E) exergonic.

100. Chemoautotrophs may capture energy from all of the following EXCEPT
    (A) CO_2.
    (B) H_2S.
    (C) NH_4^+.
    (D) NO_2^-.
    (E) All of the above might be used.

101. Of the following electron carriers of the electron transport system, which transfers protons in addition to electrons?
    (A) cytochrome a
    (B) cytochrome c
    (C) ATP synthase
    (D) coenzyme Q
    (E) cytochrome c1

102. The respiration process that results in the buildup of organic compounds in cells is known as
    (A) dehydration.
    (B) oxidation.
    (C) reduction.
    (D) anaerobiosis.
    (E) fermentation.

103. The conversion of glucose to pyruvate takes place
    (A) in the mitochondrial matrix.
    (B) in the mitochondrial cristae.
    (C) only in heterotrophic cells.
    (D) in the nucleus of eukaryotes.
    (E) none of the above

104. The process of converting pyruvate to acetyl CoA
    (A) produces one molecule of CO_2.
    (B) produces three ATP by substrate-level phosphorylation.
    (C) produces three ATP by oxidative phosphorylation.
    (D) both A and C
    (E) both A and D

105. The net CO_2 production from a single turn of the TCA cycle is
    (A) 2.
    (B) 3.
    (C) 11.
    (D) 12.
    (E) none of the above

CHAPTER 6

Photosynthesis

106. Which of the following occurs during the Calvin-Benson cycle?
    (A) synthesis of glucose
    (B) carbon fixation
    (C) photolysis
    (D) all of the above
    (E) both A and B

107. When a compound absorbs the energy of a photon of light, which of the following can occur?
    (A) It may pass off the photon to another compound.
    (B) It may hold the energy for a period of time, then release a photon of a shorter wavelength.
    (C) It may use the energy to build a molecule of water.
    (D) It may immediately release a photon with a shorter wavelength.
    (E) Either B or D may occur.

108. Pigments that can absorb light energy include
    (A) chlorophyll b.
    (B) carotenoids.
    (C) chlorophyll a.
    (D) all of the above
    (E) both A and C

109. A photon at a _____ wavelength would contain the most energy.
    (A) blue
    (B) orange
    (C) violet
    (D) red
    (E) green

110. "A photon strikes chlorophyll, the dislodged electron is used to generate a molecule of ATP, and the electron is then returned to the original molecule" describes
    (A) cyclic photophosphorylation.
    (B) noncyclic photophosphorylation.
    (C) light-independent reactions.
    (D) the Calvin-Benson cycle.
    (E) photolysis.

111. Which of the following is the smallest structure of the group?
    (A) chloroplast
    (B) thylakoid
    (C) grana
    (D) mitochondria
    (E) It is impossible to tell.

112. The Calvin-Benson cycle is also known as
    (A) cyclic photophosphorylation.
    (B) noncyclic photophosphorylation.
    (C) the light-dependent reactions.
    (D) the light-independent reactions.
    (E) both B and C

113. Cyclic photophosphorylation
    (A) occurs only in plants.
    (B) involves photosystems I and II.
    (C) occurs only in bacteria.
    (D) involves photosystems P680 and P700.
    (E) both C and D

114. The synthetic pathway that includes RuBP is known as
    (A) the Calvin-Benson cycle.
    (B) pyruvate conversion.
    (C) glycolysis.
    (D) the TCA cycle.
    (E) the chloroplast electron transport chain.

115. Molecular oxygen (O_2) is associated with
    (A) photolysis.
    (B) the mitochondrial electron transport chain.
    (C) the Calvin-Benson cycle.
    (D) all of the above
    (E) A and B only

116. Photosynthesis is an important process that
    (A) is performed by heterotrophs.
    (B) produces oxidized products.
    (C) is performed by organisms living near deep-ocean thermal vents.
    (D) only plants can perform.
    (E) uses water and CO_2 as reactants.

117. Which of the following initially traps solar energy in the process of photosynthesis?
    (A) water
    (B) chlorophyll
    (C) glucose
    (D) $NADP^+$
    (E) ATP

118. Which of the following is NOT true of phototrophs?
    (A) They capture light energy.
    (B) They can be bacteria.
    (C) They store energy as glucose.
    (D) They never function as autotrophs.
    (E) All of the above are true.

119. Which of the following is a phototroph?
    (A) a tree
    (B) a mushroom
    (C) bacteria growing on decomposing matter
    (D) an earthworm
    (E) a person

120. Oxygenic photoheterotrophs
    (A) do not exist.
    (B) produce O_2 as a waste product.
    (C) utilize CO_2 to synthesize glucose.
    (D) none of the above
    (E) both B and C

121. Which of the following is NOT a component of the thylakoid membrane?
    (A) NADP reductase
    (B) plastoquinone
    (C) ferredoxin
    (D) cytochrome b-c1 complex
    (E) all of the above

122. Bacteriorhodopsin is a(n)
    (A) green pigment.
    (B) orange pigment.
    (C) purple pigment.
    (D) oxidizing reagent.
    (E) both C and D

123. Photolysis is involved in
    (A) the cyclic pathway of ATP formation.
    (B) carotenoid pigments.
    (C) photosystem I.
    (D) the noncyclic pathway of ATP formation.
    (E) both A and C

124. _____ are organic materials that can absorb light energy.
    (A) Pigments
    (B) Lipids
    (C) Fats
    (D) both B and C
    (E) none of the above

125. Which is NOT required for cyclic photophosphorylation?
    (A) photons
    (B) pigments
    (C) electron acceptors
    (D) grana
    (E) ADP

126. Water serves as a source of _____ in chloroplasts.
    (A) oxygen waste
    (B) electrons
    (C) hydrogen ions
    (D) all of the above
    (E) both B and C

CHAPTER 7

Mitosis

127. The period of nuclear division is called
    (A) G_2 phase.
    (B) mitosis.
    (C) S phase.
    (D) G_1 phase.
    (E) cytokinesis.

128. Some cells are locked in interphase. Among these cells are those responsible for
    (A) forming clotting proteins.
    (B) immunologic memory.
    (C) memory storage in the brain.
    (D) A and B only.
    (E) B and C only.

129. The complete cell cycle is characterized by the chromosomal formula
    (A) $1n \to 2n \to 4n \to 2x2n$.
    (B) $2n \to 4n \to 8n \to 2x4n$.
    (C) $2n \to 4n \to 2x2n$.
    (D) $2n \to 4n \to 2x2n \to 4x1n$.
    (E) $4n \to 2x2n \to 2x4n$.

130. The period in which metaphase occurs is called
    (A) G_2 phase.
    (B) mitosis.
    (C) S phase.
    (D) G_1 phase.
    (E) cytokinesis.

131. When a mother cell gives rise to two identical daughter cells, the process is known as
    (A) meiosis II.
    (B) meiosis I.
    (C) mitosis.
    (D) genetic variation.
    (E) crossing over.

132. Which of the following is involved in asexual reproduction?
    (A) meiosis
    (B) transformation
    (C) mitosis
    (D) reverse transcriptase
    (E) trisomy

133. The chromosomes are moving to the opposite poles during
    (A) metaphase.
    (B) anaphase.
    (C) interphase.
    (D) prophase.
    (E) telophase.

134. Condensation and shortening of chromosomes occurs during which phase?
    (A) interphase
    (B) prophase
    (C) metaphase
    (D) anaphase
    (E) telophase

135. Cells age as a result of chromosomal degradation at the tips during the cell cycle. Stem cells can replicate indefinitely, however, because of
    (A) the presence of pristine recombinase proteins.
    (B) high levels of topoisomerase.
    (C) low levels of helicase.
    (D) the presence of telomerase.
    (E) high levels of transposase.

136. How would the loss of ribosomal function affect cell replication?
    (A) Cell replication would continue as normal but at a much slower rate.
    (B) Cell replication would cease but only after 3 to 4 cell divisions.
    (C) Cell replication would be unaffected.
    (D) Because of the energy conserved, the replication rate would actually increase.
    (E) Cell replication would cease immediately.

137. Which best describes the distribution of DNA following mitosis?
    (A) Each daughter cell gets the same amount of DNA as was in the mother cell.
    (B) Each daughter cell gets one half as much DNA as was in the mother cell.
    (C) Each daughter cell gets the same number of genes as was in the mother cell.
    (D) Each daughter cell gets one quarter as much DNA as was in the mother cell.
    (E) Both A and C are true.

138. Cytokinesis involves
    (A) the separation of the cytosols of two daughter cells.
    (B) the movement of mRNA from cell to cell.
    (C) the movement of DNA from one cell to another.
    (D) the movement of cytosol from one cell to another.
    (E) the movement of DNA via viruses from cell to cell.

139. Bacteria do not undergo mitosis. This is because they
    (A) lack a cell cycle.
    (B) are too small.
    (C) lack mitochondria.
    (D) are not able to contain the amount of energy required for the effort.
    (E) are not photosynthetic.

140. Which best describes the distribution of organelles following mitosis?
    (A) All four daughter cells receive an equal distribution.
    (B) Half of the daughter cells get more than the other half.
    (C) The distribution of organelles is equal among all daughter cells.
    (D) The distribution pattern is best described as conservative, with one half of the daughter cells getting all of the original organelles and the other getting copies.
    (E) both A and C

141. Prokaryotes replicate by
    (A) mitosis.
    (B) binary fission.
    (C) meiosis.
    (D) double fertilization.
    (E) none of the above

142. In mitosis, each daughter cell gets _____ of the mother cell's S phase DNA.
    (A) 50%
    (B) 75%
    (C) 25%
    (D) 100%
    (E) none of the above

143. Which best describes the chromosomes of eukaryotes?
    (A) They contain no proteins.
    (B) They have large amounts of protein distributed along the DNA strand.
    (C) They have small amounts of protein distributed along the DNA strand.
    (D) They have large amounts of protein located only at the ends of the DNA strand.
    (E) They have small amounts of protein located only at the ends of the DNA strand.

144. Sister chromatids
    (A) are joined at the telomers.
    (B) are inverse copies of each other.
    (C) always come from different sister cells.
    (D) are only found following double fertilization.
    (E) are joined at the centromere.

145. The amount of DNA in a fertilized ovum is _____ as the amount in a zygote.
    (A) one half as much
    (B) twice as much
    (C) the same
    (D) one quarter as much
    (E) four times as much

146. During which phase of mitosis do the chromosomes start to separate?
    (A) prophase
    (B) anaphase
    (C) metaphase
    (D) cytokinesis
    (E) telophase

147. During which phase of mitosis does chromatin become a chromosome?
    (A) prophase
    (B) anaphase
    (C) metaphase
    (D) interphase
    (E) telophase

CHAPTER 8

Meiosis

148. When a mother cell gives rise to four genetically different daughter cells, the process is known as
    (A) meiosis.
    (B) mitosis.
    (C) reverse transcription.
    (D) genetic engineering.
    (E) cloning.

149. The genetic content of gametes can best be characterized as
    (A) reductions of the mother cell.
    (B) increases of the mother cell.
    (C) identical to the mother cell.
    (D) a clone of the mother cell.
    (E) both C and D

150. Cytokinesis following meiosis I results in what distribution of DNA?
    (A) $1x2n$
    (B) $2x2n$
    (C) $1x4n$
    (D) $4x1n$
    (E) $8x1n$

151. Meiosis is characterized by the chromosomal formula
    (A) $1n \to 2n \to 4n \to 2x2n$.
    (B) $2n \to 4n \to 8n \to 2x4n$.
    (C) $2n \to 4n \to 2x2n$.
    (D) $2n \to 4n \to 2x2n \to 4x1n$.
    (E) $4n \to 2x2n \to 2x4n$.

152. During which stage do the sister chromatids separate?
    (A) prophase I
    (B) prophase II
    (C) anaphase II
    (D) anaphase I
    (E) telophase I

153. Synapsis occurs during
    (A) mitosis.
    (B) prophase I.
    (C) metaphase II.
    (D) prophase II.
    (E) anaphase I.

154. Which of the following is involved in sexual reproduction?
    (A) meiosis
    (B) transformation
    (C) mitosis
    (D) mutation
    (E) trisomy

155. Cytokinesis following meiosis II results in what distribution of DNA?
    (A) $1x2n$
    (B) $2x2n$
    (C) $1x4n$
    (D) $4x1n$
    (E) $8x1n$

156. Through meiosis
    (A) alternate forms of genes are shuffled.
    (B) parental DNA is divided and distributed to forming gametes.
    (C) the diploid chromosome number is reduced to haploid.
    (D) offspring are provided with new gene combinations.
    (E) all of the above

157. Reformation of the nuclear membrane during meiosis occurs during which phase?
    (A) metaphase I
    (B) prophase I
    (C) anaphase II
    (D) telophase I
    (D) metaphase II

158. Which of the following groups of organisms undergo meiosis?
    (A) animals
    (B) plants
    (C) fungi
    (D) all of the above
    (E) both A and B

159. Crossing over differs from cytokinesis in that
    (A) only the former actually involves the movement of DNA.
    (B) only the latter is found in both mitosis and meiosis.
    (C) only the former involves the centromere.
    (D) only the former is seen in both prokaryotes and eukaryotes.
    (E) the former involves DNA, while the latter involves RNA.

160. All of the following involve genetic variation. Which is the exception?
    (A) sexual reproduction
    (B) crossing over
    (C) binary fission
    (D) random alignment of chromosomes during meiosis
    (E) recombination of alleles

161. When plant gametes fuse, the process is known as
    (A) fertilization.
    (B) gametogenesis.
    (C) fission.
    (D) meiosis.
    (E) morphogenesis.

162. What occurs during prophase II?
    (A) The nuclear envelopes disappear.
    (B) The chromosomes migrate toward the cellular poles.
    (C) The chromosomes immobilize along the cellular equator.
    (D) The spindle apparatus is assembled.
    (E) The chromosomes revert to chromatin.

163. In the formation of gametes from a mother cell that is heterozygous for gene D
    (A) all of the resulting gametes will be heterozygous as well.
    (B) all of the resulting gametes will be homozygous.
    (C) each resulting gamete will have an equal chance of receiving D or d.
    (D) each resulting gamete will have an unequal chance of receiving D only.
    (E) each resulting gamete will have an unequal chance of receiving d only.

164. Sperm cells are the end product of
    (A) spermatogenesis.
    (B) gametogenesis.
    (C) fertilization.
    (D) all of the above
    (E) both A and B

165. Sexual recombination in prokaryotes
    (A) produces smaller daughter cells.
    (B) produces larger daughter cells.
    (C) is a result of binary fission rather than meiosis.
    (D) commonly results in the formation of new species.
    (E) does not occur.

166. The primary difference between a zygote and an embryo is that
    (A) only the latter is seen in plants.
    (B) only the former is a single cell.
    (C) only the latter is diploid.
    (D) only the former is haploid.
    (E) only the latter is seen in animals.

167. The proteins required for changing the genetic content of a cell from diploid to tetraploid are manufactured during which phase?
    (A) anaphase I
    (B) prophase I
    (C) G_2 phase
    (D) metaphase I
    (E) G_0 phase

168. The proteins required for meiosis are manufactured during which phase?
    (A) anaphase I
    (B) prophase I
    (C) G_2 phase
    (D) metaphase I
    (E) G_0 phase

CHAPTER 9

Inheritance

169. _____ is a blood disorder where clotting is dysfunctional.
    (A) Tay-Sachs
    (B) Cystic fibrosis
    (C) Hemophilia
    (D) Sickle cell disease
    (E) none of the above

170. A father with type A blood and a mother with type B blood will always have children with
    (A) blood type A.
    (B) blood type B.
    (C) blood type AB.
    (D) blood type O.
    (E) none of the above

171. _____ is caused by excess mucus production in the lungs.
    (A) Tay-Sachs
    (B) Cystic fibrosis
    (C) Hemophilia
    (D) Sickle cell disease
    (E) none of the above

172. The genetic defect that was associated with nineteenth- and twentieth-century European royalty was
    (A) Tay-Sachs.
    (B) cystic fibrosis.
    (C) hemophilia.
    (D) sickle cell disease.
    (E) none of the above

173. A genotype of *RR* would indicate
    (A) heterozygous on autosomes.
    (B) heterozygous on male sex chromosomes.
    (C) homozygous dominant on autosomes.
    (D) homozygous recessive on autosomes.
    (E) homozygous dominant on male sex chromosomes.

174. The alleles associated with the Rh blood groups are
    (A) A, B, and AB.
    (B) A, B, and O.
    (C) *R* and *r*.
    (D) Rh^+ and Rh^-.
    (E) A, B, AB, and O.

175. A father with type A blood and a mother with type A blood
    (A) will always have children with blood type A.
    (B) will always have children with blood type B.
    (C) may have children with blood type AB.
    (D) may have children with blood type O.
    (E) none of the above

176. _____ is caused by a buildup of a single amino acid producing mental retardation.
    (A) Tay-Sachs
    (B) Cystic fibrosis
    (C) Hemophilia
    (D) Sickle cell disease
    (E) none of the above

177. If *R* is dominant to *r*, the offspring of the cross of *RR* with *rr* will
    (A) be homozygous.
    (B) display the same phenotype as the *RR* parent.
    (C) display the same phenotype as the *rr* parent.
    (D) have the same genotype as the *RR* parent.
    (E) have the same genotype as the *rr* parent.

178. The genetic defect associated with a single amino acid change in hemoglobin is
    (A) Tay-Sachs.
    (B) cystic fibrosis.
    (C) hemophilia.
    (D) sickle cell disease.
    (E) none of the above

179. A genotype of R^- would indicate
    (A) heterozygous on autosomes.
    (B) heterozygous on male sex chromosomes.
    (C) homozygous dominant on autosomes.
    (D) homozygous recessive on autosomes.
    (E) hemizygous on male sex chromosomes.

180. _____ is a blood disorder where low oxygen levels may induce a crisis.
    (A) Tay-Sachs
    (B) Cystic fibrosis
    (C) Hemophilia
    (D) Sickle cell disease
    (E) none of the above

181. The alleles associated with the ABO blood groups are
    (A) A, B, and AB.
    (B) A, B, and O.
    (C) R and r.
    (D) Rh^+ and Rh^-.
    (E) A, B, AB, and O.

182. To express a recessive X-linked trait,
    (A) a female must be heterozygous.
    (B) a female must be homozygous.
    (C) a male must be heterozygous.
    (D) a male must be heterozygous.
    (E) both A and C

183. _____ is a disease where excess lipids cause nerve dysfunction.
    (A) Tay-Sachs
    (B) Cystic fibrosis
    (C) Hemophilia
    (D) Sickle cell disease
    (E) none of the above

184. A pair of homologous chromosomes may differ from other chromosomes in terms of
    (A) size.
    (B) banding patterns.
    (C) the alleles they carry.
    (D) position of the centromere.
    (E) all of the above

185. A genotype of *Rr* would indicate
    (A) heterozygous on autosomes.
    (B) heterozygous on male sex chromosomes.
    (C) homozygous dominant on autosomes.
    (D) homozygous recessive on autosomes.
    (E) hemizygous on male sex chromosomes.

186. Which is NOT true of human chromosomes?
    (A) The haploid number is 23.
    (B) The diploid number is 46.
    (C) There are 23 pairs of chromosomes.
    (D) Human gametes end up with two of each type of 23 chromosomes.
    (E) Human gametes end up with one of each type of 23 chromosomes.

187. The genetic defect that can be affected by the sweetener aspartame (Equal®) is
    (A) Tay-Sachs.
    (B) cystic fibrosis.
    (C) hemophilia.
    (D) sickle cell disease.
    (E) none of the above

188. Sickle cell anemia
    (A) causes cramping and death.
    (B) is precipitated by high oxygen levels.
    (C) is autosomal dominant.
    (D) all of the above
    (E) both A and B

189. Considering crossing possibilities for three genes simultaneously would require the construction of a Punnett square with _____ cells.
    (A) 4
    (B) 16
    (C) 32
    (D) 64
    (E) 128

CHAPTER 10

DNA Replication and Repair

190. DNA replication is
    (A) conservative.
    (B) dispersive.
    (C) semiconservative.
    (D) irregular.
    (E) none of the above

191. UV light–treated bacteria would most likely have DNA damage in the form of
    (A) analog incorporation.
    (B) pyrimidine dimer formation.
    (C) intercalation of the bases.
    (D) direct transition of the bases.
    (E) deamination.

192. In 1928, experiments with *Streptococcus pneumoniae* suggested that
    (A) bacteria do not incorporate the DNA of other bacteria.
    (B) rough bacteria spontaneously convert to the smooth type.
    (C) DNA was conclusively the only molecule associated with heredity.
    (D) heat-killed bacteria could somehow "transform" live bacteria.
    (E) all of the above

193. A change in a nucleotide sequence that results in the addition or deletion of a single nucleotide and largely changes the amino acid sequence of the resulting peptide is known as a
    (A) nonsense mutation.
    (B) silent mutation.
    (C) missense mutation.
    (D) frameshift mutation.
    (E) none of the above

194. The material responsible for melting short segments of DNA is
    (A) single-stranded DNA binding proteins.
    (B) helicase.
    (C) topoisomerase.
    (D) DNA ligase.
    (E) none of the above

195. An example of a purine is
    (A) thymine.
    (B) guanine.
    (C) acridine.
    (D) cytosine.
    (E) uracil.

196. In a well-known experimental series in 1952, ^{35}S was added to viruses replicating within its bacterial host. The new virions were carefully isolated and used to infect fresh bacterial cells in the absence of radioisotopes. Where would you expect to find the ^{35}S radioisotope?
    (A) in the phage proteins left outside the newly infected cells
    (B) in the bacterial proteins
    (C) in the bacterial DNA
    (D) in the nucleic acid of the new virions
    (E) in the protein coat of the new virions

197. _____ covalently links breaks in the sugar-phosphate backbone.
    (A) Helicase
    (B) Gyrase
    (C) Ligase
    (D) Primase
    (E) none of the above

198. In the following gene sequence, the original sequence was A-B-C-D-E-F-G-H. It now reads A-B-C-D-G-F-E-H. This type of mutation is a(n)
    (A) nonsense.
    (B) inversion.
    (C) translocation.
    (D) duplication.
    (E) point mutation.

DNA Replication and Repair

199. DNA isolated from cow liver cells contains 28% A. What percentage will be C?
    (A) 14%
    (B) 22%
    (C) 28%
    (D) 36%
    (E) 56%

200. _____ unwinds double-stranded DNA.
    (A) Helicase
    (B) Gyrase
    (C) Ligase
    (D) Primase
    (E) none of the above

201. The replisome is associated with
    (A) transcription.
    (B) replication.
    (C) translation.
    (D) lagging strand DNA synthesis.
    (E) both B and D

202. _____ nicks DNA ahead of the replication fork to relax supercoiling in the DNA.
    (A) Helicase
    (B) Gyrase
    (C) Ligase
    (D) Primase
    (E) none of the above

203. The "central dogma" of retroviruses is most accurately described as
    (A) DNA → protein.
    (B) RNA → protein.
    (C) DNA → RNA → protein.
    (D) protein → RNA → DNA → protein.
    (E) RNA → DNA → RNA → protein.

204. The method used by bacteria to replicate their DNA during conjugation is
    (A) alpha mode replication.
    (B) theta mode replication.
    (C) rolling circle replication.
    (D) ribosomal replication.
    (E) both A and C

205. The most vigorous and aggressive form of DNA repair is
   (A) recombination repair.
   (B) excision repair.
   (C) SOS repair.
   (D) exonuclease proofreading.
   (E) photoreactivation.

206. Which of the following wavelengths contains enough energy to damage DNA?
   (A) microwaves
   (B) long-wave radio waves
   (C) short-wave radio waves
   (D) visible light
   (E) none of the above

207. If I isolated a gene identified as *recA*, then I would be studying
   (A) transcription.
   (B) insertion sequences.
   (C) DNA repair mechanisms.
   (D) transposons.
   (E) translation.

208. DNA polymerase
   (A) reads template DNA 3′ to 5′.
   (B) reads template DNA 5′ to 3′.
   (C) synthesizes DNA 3′ to 5′.
   (D) both B and C
   (E) none of the above

209. The enzyme involved in the process called replication is
   (A) DNA-dependent DNA polymerase.
   (B) DNA-dependent RNA polymerase.
   (C) RNA-dependent RNA polymerase.
   (D) RNA-dependent DNA polymerase.
   (E) both A and B

210. Which of the following enzymes is NOT associated with DNA replication?
   (A) DNA ligase
   (B) RNA primase
   (C) ATP synthase
   (D) DNA polymerase
   (E) topoisomerase

CHAPTER 11

RNA Transcription

211. A chemical substance that would interfere with DNA replication and/or transcription by physically changing the structure of one of the DNA bases already in the genome would be called

(A) a mutagen.
(B) an intercalating agent.
(C) a base modifier.
(D) both A and B
(E) both A and C

212. Which of the following best describes cDNA?

(A) It is produced directly by removing the exons from hnRNA.
(B) It is produced directly by removing the introns from hnRNA.
(C) It is produced by back-sequencing from the expressed protein.
(D) It is produced from mRNA by reverse transcriptase.
(E) It is produced from DNA by a restriction endonuclease.

213. The enzyme involved in the process called transcription is

(A) DNA-dependent DNA polymerase.
(B) DNA-dependent RNA polymerase.
(C) RNA-dependent RNA polymerase.
(D) RNA-dependent DNA polymerase.
(E) none of the above

214. RNA can be used to produce DNA via the action of

(A) DNA-dependent DNA polymerase.
(B) DNA-dependent RNA polymerase.
(C) reverse transcriptase.
(D) ligase.
(E) restriction endonuclease.

215. Repressors bind at which region of an operon?
    (A) the operator
    (B) *lacZ*
    (C) the structural genes
    (D) the regulator genes
    (E) none of the above

216. The Pribnow box is
    (A) a region of RNA that codes for the final enzyme.
    (B) a region of DNA that is removed during posttranscriptional modification.
    (C) a region of RNA that is removed during posttranscriptional modification.
    (D) a conserved sequence within the promoter region.
    (E) none of the above

217. The component of the RNA polymerase holoenzyme that determines specificity of the precise binding site in prokaryotes is the
    (A) β' subunit.
    (B) σ subunit.
    (C) α subunit.
    (D) β' subunit.
    (E) The entire enzyme determines specificity.

218. RNA polymerase binds at which region of an operon?
    (A) the operator
    (B) *lacZ*
    (C) the structural genes
    (D) the regulator genes
    (E) none of the above

219. Which of the following is the immediate product of prokaryotic transcription?
    (A) hnRNA
    (B) ribosomes
    (C) DNA
    (D) mRNA
    (E) none of the above

220. Which of the following is/are posttranscriptional modification(s) of hnRNA done by eukaryotes?
    (A) addition of a 3' poly-A tail
    (B) addition of a 5'-7-methylguanosine cap
    (C) removal of introns
    (D) both A and B
    (E) A, B, and C

221. Transcription is the process of
    (A) making DNA from tRNA sequences.
    (B) making DNA from DNA sequences.
    (C) making proteins from mRNA codes.
    (D) making proteins from DNA codes.
    (E) none of the above

222. A section of nucleic acid composed of both purines and pyrimidines could
    (A) be double stranded.
    (B) be single stranded.
    (C) have a backbone containing ribose.
    (D) all of the above
    (E) none of the above

223. Which of the following best describes RNA polymerase?
    (A) reads DNA and makes DNA
    (B) reads DNA and makes mRNA
    (C) reads proteins and makes codons
    (D) reads mRNA and makes proteins
    (E) both A and B

224. A mutation in a promoter region could result in which of the following changes?
    (A) changing an operon from inducible to constitutive
    (B) changing an operon from inducible to permanently repressed
    (C) changing the affinity of binding of DNA polymerase
    (D) all of the above
    (E) both B and C

225. Which of the following is a process done by prokaryotes on their mRNA?
    (A) addition of a 5′ ATP cap
    (B) addition of a 3′ poly-A tail
    (C) removal of introns
    (D) removal of exons
    (E) both B and C

226. The *trp* operon is used by bacteria for
    (A) utilization of tryptophan.
    (B) production of tryptophan.
    (C) synthesis of CAP.
    (D) production of tRNA.
    (E) none of the above

227. Which of the following is a posttranscriptional modification of mRNA done by eukaryotes?
    (A) addition of a 3′ poly-A tail
    (B) addition of a 5′-7-methylguanosine cap
    (C) removal of introns
    (D) removal of exons
    (E) A, B, and C

228. A large segment of nucleic acid that is entirely lacking in thymidine would most likely
    (A) be double stranded.
    (B) be single stranded.
    (C) have a backbone containing ribose.
    (D) none of the above
    (E) both B and C

229. The glucose operon is an example of a(n)
    (A) constitutive operon.
    (B) inducible operon.
    (C) repressible operon.
    (D) all of the above
    (E) none of the above

230. The DNA sequence TATAAT is associated with
    (A) the binding site of DNA polymerase.
    (B) the binding site of RNA polymerase.
    (C) the binding site of a repressor.
    (D) the binding site of an inhibitor.
    (E) both C and D

231. A mutation within the Pribnow box of a critical gene would most likely result in
    (A) a greatly increased growth rate.
    (B) the silencing of that gene.
    (C) the buildup of cAMP in the cell.
    (D) the death of the cell.
    (E) both B and D

CHAPTER 12

Translation

232. In which of the following organisms would you find the 70S ribosome?
    (A) trees
    (B) humans
    (C) bacteria
    (D) all of the above
    (E) A and B only

233. Inactive ribosomes are composed of
    (A) ribosomal RNA.
    (B) transfer RNA.
    (C) proteins.
    (D) cDNA
    (E) both A and C

234. When is energy expended during protein synthesis?
    (A) only when the protein is released from the ribosome
    (B) only when the initial amino acid is brought into the ribosome
    (C) only every third amino acid addition
    (D) only when a START codon is encountered
    (E) only at the addition of every amino acid to the expanding peptide

235. The locations known as the A and P sites are found in
    (A) the nucleus.
    (B) mitochondria.
    (C) ribosomes.
    (D) centromeres.
    (E) none of the above

236. The enzyme involved in the process called translation is
    (A) DNA-dependent DNA polymerase.
    (B) DNA-dependent RNA polymerase.
    (C) RNA-dependent RNA polymerase.
    (D) RNA-dependent DNA polymerase.
    (E) none of the above

237. If the DNA triplets were ATG-CGT and were used to synthesize mRNA, the anticodons would be
    (A) AUG-CGU.
    (B) ATG-CGT.
    (C) UAC-GCA.
    (D) UAG-CGU.
    (E) none of the above

238. Which of the following would you find within the ribosome as a peptide is being synthesized?
    (A) mRNA
    (B) rRNA
    (C) tRNA
    (D) all of the above
    (E) both A and B

239. Which of the following carries amino acids to the ribosome?
    (A) tRNA
    (B) rRNA
    (C) DNA
    (D) mRNA
    (E) both A and B

240. Which, if any, of the following nucleic acid sequences represents a start codon?
    (A) UGA
    (B) AUG
    (C) UAA
    (D) UAG
    (E) none of the above

241. The enzyme β-galactosidase is a key enzyme associated with
    (A) tryptophan hydrolysis.
    (B) glucose utilization.
    (C) lactose synthesis.
    (D) lactose utilization.
    (E) tryptophan synthesis.

242. Control of protein synthesis can be exerted by a _____ protein that binds to the DNA region that controls the expression of structural genes.
    (A) stimulator
    (B) detractor
    (C) operator
    (D) repressor
    (E) inhibitor

243. The three phases of translation are initiation, termination, and
    (A) continuation.
    (B) elongation.
    (C) release.
    (D) ATP activation.
    (E) none of the above

244. Which of the following can be found within the 30S ribosome subunit?
    (A) tRNA
    (B) rRNA
    (C) DNA
    (D) mRNA
    (E) none of the above

245. Which of the following is NOT directly associated with ribosomes?
    (A) tRNA
    (B) rRNA
    (C) DNA
    (D) mRNA
    (E) both A and B

246. What happens to a molecule of rRNA after it is transcribed?
    (A) It immediately associates with ribosomal proteins.
    (B) It is transported back into the nucleus for ribosomal subunit assembly.
    (C) It is immediately degraded.
    (D) It is used as a template for ribosomal assembly.
    (E) It immediately associates with tRNAs.

247. Within a functioning ribosome, which is most numerous?
    (A) tRNAs
    (B) rRNAs
    (C) ribosomal proteins
    (D) mRNA
    (E) assembling amino acids

248. There are _____ codons within the genetic code.
    (A) 3
    (B) 4
    (C) 16
    (D) 32
    (E) 64

249. Of the possible nucleotide combinations within the genetic code, how many are directly associated with the termination of protein synthesis?
    (A) none
    (B) 3
    (C) 4
    (D) 8
    (E) 12

250. Why is energy required for the process of translation?
    (A) Because it is necessary to form the peptide bond between amino acids.
    (B) Because it is necessary to dissociate the ribosomal subunits after termination.
    (C) Because it is used to move the ribosome along the mRNA transcript.
    (D) Because it is necessary to hybridize the tRNAs to the mRNA.
    (E) Energy is not required for translation.

251. The portion of any eukaryotic RNA transcript used for translation is composed of
    (A) anticodons.
    (B) exons.
    (C) introns.
    (D) leader sequences.
    (E) inversions.

252. What is a polysome?
    (A) a ribosome that is associated only with mitosis
    (B) a ribosome found only in mitochondria
    (C) a series of ribosomes associated only with gametogenesis
    (D) a ribosome that can produce a protein starting at the end of the mRNA molecule farthest from the leader
    (E) none of the above

CHAPTER 13

Genetic Engineering

253. The purpose of a restriction endonuclease is to
    (A) introduce DNA into a host cell.
    (B) join DNA fragments.
    (C) fragment DNA.
    (D) amplify DNA fragments.
    (E) all of the above

254. Which of the following best describes the primers used in the PCR technique?
    (A) One, two, or three different sequences may be used with equal efficiency at the same time.
    (B) They are generally 18 to 25 bases in length.
    (C) They are generally computer designed.
    (D) all of the above
    (E) both B and C

255. The term "pBR322" refers to a genetic engineering tool classified as a
    (A) plasmid.
    (B) restriction endonuclease.
    (C) transposon.
    (D) restriction fragment.
    (E) phage.

256. *Taq*, when referring to the PCR technique, identifies
    (A) a specific restriction endonuclease.
    (B) a specific enzyme cofactor.
    (C) a specific source of DNA polymerase.
    (D) the original source of the required primer pairs.
    (E) none of the above

257. A gene gun is used to
    (A) remove selected genes from bacterial cells.
    (B) remove selected genes from eukaryotic cells.
    (C) insert modified ribosomes into bacterial cells.
    (D) insert selected genes into eukaryotic cells.
    (E) produce knockout mice with missing genes.

258. If I wanted to get specific DNA into a recipient cell, I could use the technique called
    (A) microinjection.
    (B) electroporation.
    (C) cosmid transduction.
    (D) plasmid transformation.
    (E) all of the above

259. A northern blot is the electrophoresis and transfer onto a membrane of which of the following compounds?
    (A) protein
    (B) DNA
    (C) RNA
    (D) lipids
    (E) none of the above

260. Autoradiographs can be used to
    (A) identify the location of specific proteins in a western blot.
    (B) identify the location of specific proteins in a Southern blot.
    (C) locate genes on whole chromosomes.
    (D) split DNA, producing sticky ends.
    (E) update gene libraries.

261. An oligonucleotide probe can be labeled with _____ to detect _____.
    (A) an enzyme; a protein
    (B) radioactive phosphorus; DNA fragments
    (C) radioactive phosphorus; a protein
    (D) all of the above
    (E) both B and C

Genetic Engineering

262. Separation of DNA fragments within an agarose matrix by applying an electrical current is
    (A) a western blot.
    (B) a Southern blot.
    (C) a northern blot.
    (D) electrophoresis.
    (E) none of the above

263. What is "golden rice"?
    (A) an especially tasty variety of rice from China
    (B) a GMO containing carotene
    (C) a plant produced by crossing carrots and rice
    (D) a very expensive strain of rice
    (E) an infected rice grain identified by its yellow color

264. Which of the following is NOT a "natural" means of transferring DNA in prokaryotes?
    (A) conjugation
    (B) transduction
    (C) electroporation
    (D) transformation
    (E) infection

265. Restriction endonucleases
    (A) protect bacteria from phage infection.
    (B) cut double-stranded DNA.
    (C) produce sticky ends.
    (D) all of the above
    (E) both B and C

266. _____ is a technique used for the detection of specific separated DNA sequences.
    (A) Southern blot
    (B) Eastern blot
    (C) Western blot
    (D) Northern blot
    (E) SDS-PAGE

267. A standard material used for the electrophoresis of protein is
    (A) polyacrylamide.
    (B) agarose.
    (C) nitrocellulose.
    (D) *Taq* polymerase.
    (E) restriction endonuclease.

268. Which of the following is an example of a genetically engineered organism?
    (A) a bacterium containing the human gene for insulin
    (B) a corn plant crossbred with wheat to increase yield
    (C) a tulip infected with a virus that produces a variegated color pattern
    (D) a bacterium containing a resistance plasmid
    (E) none of the above

269. The technique used to amplify one copy of DNA to millions of copies is
    (A) electrophoresis.
    (B) polymerase chain reaction.
    (C) western blot.
    (D) Southern blot.
    (E) none of the above

270. The technique referred to as RFLP
    (A) involves the production of hormones.
    (B) is a form of gene therapy.
    (C) inserts the BT gene into plants.
    (D) was used to make "golden rice."
    (E) none of the above

271. _____ is a technique used for the detection of specific separated protein sequences.
    (A) Southern blot
    (B) Eastern blot
    (C) Western blot
    (D) Northern blot
    (E) SDS-PAGE

272. What is a reporter gene?
    (A) a DNA sequence found only in mammals
    (B) a gene that codes for splice sites within a plasmid
    (C) a gene that codes for splice sites within a bacterial genome
    (D) a gene that produces a readily identifiable product
    (E) a gene that produces a commercially valuable product

273. A certain type of bacterial plasmid produces products that protect the host cell from harmful chemical attacks. This is known as a(n) ____ plasmid.
    (A) R
    (B) F
    (C) F+
    (D) constructed
    (E) symbiotic

CHAPTER 14

Origin of Life

274. The earth's atmosphere before the existence of life is thought to have lacked
    - (A) inert gases.
    - (B) molecular nitrogen.
    - (C) molecular oxygen.
    - (D) water vapor.
    - (E) molecular hydrogen.

275. Life in its most primitive form is thought to have originated on the earth about _____ years ago.
    - (A) 1.5 million
    - (B) 3.5 million
    - (C) 6 billion
    - (D) 3.5 billion
    - (E) 4.5 billion

276. Life on the earth is possible primarily because of
    - (A) the presence of liquid water.
    - (B) the presence of molecular oxygen.
    - (C) continental drift.
    - (D) lack of competition from other planets.
    - (E) the presence of silica-containing surfaces.

277. Which is thought to be the first nucleic acid produced abiotically?
    - (A) rDNA
    - (B) tRNA
    - (C) RNA
    - (D) mtDNA
    - (E) DNA

278. Radioactive elements undergo radioactive decay, which is
    (A) different for every isotope.
    (B) based on the nuclear instability within the ratio of neutrons to protons.
    (C) independent of surrounding environmental conditions.
    (D) constant for any given isotope.
    (E) all of the above

279. In order for proteins to be abiotically produced in a primitive atmosphere, what must already be present?
    (A) nitrogen dioxide
    (B) carbon dioxide
    (C) a DNA template
    (D) amino acids
    (E) an RNA template

280. What is thought to have been essential for the abiotic synthesis of proteins?
    (A) layers of dried mud
    (B) hollow pockets in solidified basalt
    (C) layers of crystallized mica
    (D) clay deposits in tidal pools
    (E) clay crystals

281. _____ were the researchers who first proposed that simple organic molecules could be constructed without organic processes.
    (A) Darwin and Wallace
    (B) Crick and Watson
    (C) Haldane and Oparin
    (D) Huxley and Hutton
    (E) Miller and Fox

282. Eukaryotes are thought to have developed from prokaryotic progenitors about _____ billion years ago.
    (A) 5.5
    (B) 2.5
    (C) 1.5
    (D) 3.5
    (E) 4.5

283. The endosymbiotic theory can explain all of the following except
    (A) the DNA found in chloroplasts.
    (B) the formation of the nuclear membranes.
    (C) the differences between prokaryotic and eukaryotic ribosomes.
    (D) the DNA found in mitochondria.
    (E) the nuclear genes necessary for mitochondrial function.

284. Sediments found in the earth's crust consist of _____ layers.
    (A) 2
    (B) 8
    (C) 3
    (D) 5
    (E) 4

285. The source of energy first used to simulate abiotic construction of organic molecules was
    (A) gamma radiation.
    (B) ultraviolet light.
    (C) infrared light.
    (D) thermal energy.
    (E) simulated lightning.

286. Most of the best Precambrian fossils are found in which location?
    (A) Australia
    (B) North America
    (C) South America
    (D) Asia
    (E) Africa

287. _____ is the period identified as the "Age of Reptiles."
    (A) Permian
    (B) Precambrian
    (C) Cenozoic
    (D) Devonian
    (E) Mesozoic

288. The most recent of the generally recognized geologic eras is known as the _____ era.
    (A) Silurian
    (B) Paleozoic
    (C) Precambrian
    (D) Mesozoic
    (E) Cenozoic

289. If envisioning the geologic history of the earth as an analog clock face with the origin of life occurring at midnight and the present day at noon, then the Paleozoic period would start at
    (A) 10:00 A.M.
    (B) 2:00 A.M.
    (C) 5:00 A.M.
    (D) 11:40 A.M.
    (E) 8:30 A.M.

290. The one major animal phylum NOT found within fossils dated to the Cambrian period is
    (A) Mollusca.
    (B) Chordata.
    (C) Echinodermata.
    (D) Annelida.
    (E) none of the above

291. About _____ years ago the land mass called Pangea is thought to have broken up into the current continents.
    (A) 2 billion
    (B) 50 million
    (C) 700 million
    (D) 200 million
    (E) 3.5 billion

292. The least inclusive of the taxonomic divisions routinely used to identify an organism is
    (A) kingdom.
    (B) species.
    (C) family.
    (D) order.
    (E) genus.

293. During the most recent ice age, glaciers extended as far south as
    (A) Ontario.
    (B) Texas.
    (C) the Ohio River.
    (D) central Mexico.
    (E) southern Kentucky.

294. There is geologic evidence to support the idea that a mass biologic extinction due to a meteorite's collision with the earth ended the _____ period.
    (A) Jurassic
    (B) Pleistocene
    (C) Oligocene
    (D) Cretaceous
    (E) Precambrian

CHAPTER 15

Evolutionary Mechanisms and Speciation

295. The inner ear of mammals appears to have
- (A) originated from the jawbones of reptiles.
- (B) formed from no previously observed structure.
- (C) developed as an invagination of the outer ear.
- (D) originated as a fracture of skull bones.
- (E) initially been an ossified gland.

296. The best definition of a mutation is
- (A) any change in the DNA template.
- (B) any modification of morphology.
- (C) when one species has changed into a separate species.
- (D) when the genetic code changes.
- (E) when, during translation, there is an mRNA:tRNA mismatch.

297. Dinosaurs identified as archosaurs can be found today only as
- (A) snakes.
- (B) lizards.
- (C) Gila monsters.
- (D) crocodiles.
- (E) tortoises.

298. The process of adaptive radiation is best exemplified by the
- (A) invasive species of the Americas.
- (B) Madagascar monkeys.
- (C) platypuses of Australia.
- (D) bee-eaters of Africa.
- (E) finches of the Galapagos.

299. What is the best measure of the speciation process?
 (A) organic modifications
 (B) reproductive success
 (C) migratory distances
 (D) genetic identity
 (E) host-parasite interactions

300. The feathers of birds are thought to have derived from the
 (A) hair of mammals.
 (B) tests of amoebas.
 (C) scales of reptiles.
 (D) scales of fish.
 (E) cilia of protozoans.

301. How might similar organisms living in different habitats but within the same range be isolated?
 (A) behaviorally
 (B) mechanically
 (C) ecologically
 (D) temporally
 (E) biochemically

302. Among the following, which has changed the least since their first appearance within the fossil record?
 (A) mosses
 (B) humans
 (C) reptiles
 (D) angiosperms
 (E) fish

303. A _____ is considered the earliest progenitor to all of the others within the following group.
 (A) human
 (B) mouse
 (C) mole
 (D) rat
 (E) shrew

304. Which of the following was used by Darwin to demonstrate evolutionary adaptation by finches for specialized niches on the Galapagos?
    (A) wing size
    (B) beaks
    (C) feet
    (D) color
    (E) weight

305. A common thought among scientists is that dinosaurs became extinct because
    (A) continental drift produced mountain ranges that prevented migrations.
    (B) of a period of global warming.
    (C) of the appearance of new highly infectious agents.
    (D) of a precipitous drop in global temperatures ("nuclear winter") brought about by a meteorite impact.
    (E) faster evolving animals became aggressively more competitive.

306. It is thought that humans have led the way in a major period of extinction because of
    (A) excessive competition.
    (B) fossil fuel consumption.
    (C) toxic waste generation.
    (D) modification of the world's habitats.
    (E) overhunting.

307. Which of the following has proven to be the most successful group of organisms?
    (A) mammals
    (B) fish
    (C) insects
    (D) bivalves
    (E) bacteria

308. _____ is the most common mechanism of behavioral isolation in birds.
    (A) Song pattern
    (B) Feeding method
    (C) Nesting preference
    (D) Beak size
    (E) Territorial distribution

309. Phenotypic similarities between organisms
    (A) are the only mechanism that can be used for classifying fossils.
    (B) are sufficient for full species identification.
    (C) are always linked to the genetic content of both.
    (D) never depend on the environment.
    (E) are key to reproductive success.

310. Prezygotic isolating mechanisms include all of the following except
    (A) behavior.
    (B) hybridization.
    (C) asynchronous reproductive cycles.
    (D) geographical separation.
    (E) pheromones.

311. Gene flow can occur only within which of the following?
    (A) ecosystem
    (B) community
    (C) genus
    (D) species
    (E) population

312. _____ are thought to be the first forms of mammals.
    (A) Insect-eaters
    (B) Ungulates
    (C) Rodents
    (D) Echinoderms
    (E) Dolphins

313. Speciation can result from new combinations of genes, which in turn may be produced by
    (A) crossing over.
    (B) sexual reproduction.
    (C) immigration.
    (D) mutation.
    (E) all of the above

314. Who else besides Darwin is best associated with the theory of evolution?
    (A) Oparin
    (B) Malthus
    (C) Wallace
    (D) Huxley
    (E) Lamarck

315. _____ is observed when a large percentage of a gene pool is reduced by the sudden elimination of a number of phenotypes.
    (A) The founder principle
    (B) The bottleneck effect
    (C) Genetic isolation
    (D) Natural selection
    (E) Genetic drift

CHAPTER 16

Viruses

316. What characteristics do all cells have that all viruses do not?
  - (A) the presence of metabolic mechanisms
  - (B) the presence of both DNA and RNA
  - (C) the ability to replicate
  - (D) all of the above
  - (E) both A and B

317. What do multipartite viruses and viroids have in common?
  - (A) They are all negative sense viruses.
  - (B) They are all associated with plants.
  - (C) They are all enveloped.
  - (D) They all contain their own ribosomes.
  - (E) They are all helical in morphology.

318. For a virus, what is the primary function of a peplomer?
  - (A) identification and taxonomy
  - (B) attachment to the host cell
  - (C) motility
  - (D) injection of the genome into the host cell
  - (E) both B and D

319. A(n) _____ viral genome can be translated by reading the template in both the 5′ to 3′ and 3′ to 5′ directions.
  - (A) ambisense
  - (B) positive sense
  - (C) negative sense
  - (D) Baltimore class 6 or 7
  - (E) none of the above

320. A lipid bilayer that surrounds some viruses is called a(n)
   (A) peplomer.
   (B) capsomer.
   (C) envelope.
   (D) nucleocapsid.
   (E) none of the above

321. A viral infection that remains inapparent and produces no disease for periods of up to several years but later causes problems would be classified as a(n) _____ infection.
   (A) latent
   (B) acute
   (C) transforming
   (D) massive
   (E) false

322. A tissue culture cell line that dies out soon after culture from the original tissue would probably be classified as a _____ cell line.
   (A) transformed
   (B) primary
   (C) diploid
   (D) discontinuous
   (E) both B and C

323. Of the following, which is NOT descriptive of all viruses?
   (A) are crystallizable
   (B) have a lipid bilayer cell membrane
   (C) lack metabolic activity
   (D) demonstrate an eclipse period during replication
   (E) All of the above describe viruses.

324. An individual protein component of a viral protein coat is called a
   (A) capsid.
   (B) capsomer.
   (C) nucleocapsid.
   (D) envelope.
   (E) none of the above

325. Of the following, which would most likely propagate in *E. coli*?
    (A) rabies virus
    (B) herpes simplex
    (C) φx174
    (D) *Chlamydia trachomatis*
    (E) orf virus

326. The genome of a parvovirus is
    (A) double-stranded DNA.
    (B) double-stranded RNA.
    (C) single-stranded RNA.
    (D) a protein.
    (E) single-stranded DNA.

327. When a virus takes over a host cell, then the host cell manufactures
    (A) viral proteins.
    (B) viral nucleic acids.
    (C) cellular proteins.
    (D) both A and B
    (E) both B and C

328. Viruses can infect
    (A) animals.
    (B) plants.
    (C) bacteria.
    (D) all of the above
    (E) both A and C

329. Cervical cancer is primarily caused by
    (A) a herpes virus.
    (B) an influenza virus.
    (C) a prion.
    (D) HPV.
    (E) none of the above

330. At a minimum, a virus consists of
    (A) a protein coat.
    (B) a plasma membrane.
    (C) a single nucleic acid.
    (D) all of the above
    (E) both A and C

331. Which disease is NOT caused by a virus?
    (A) smallpox
    (B) polio
    (C) syphilis
    (D) influenza
    (E) herpes

332. The first step in the process of viral replication is
    (A) penetration of the host cell membrane.
    (B) assembly.
    (C) attachment.
    (D) takeover of host cellular machinery.
    (E) release.

333. The type of agent that causes AIDS is classified as a(n)
    (A) bacterium.
    (B) viroid.
    (C) prion.
    (D) protozoan.
    (E) none of the above

334. A prion is
    (A) a virus that infects bacteria.
    (B) an infectious protein.
    (C) a virus related to influenza.
    (D) a vaccine that protects against bacterial infections.
    (E) none of the above

335. The process involving a sudden, almost explosive release of viruses from a host cell is called
    (A) lysogeny.
    (B) lysis.
    (C) a prion.
    (D) assembly.
    (E) none of the above

336. A disease caused by an agent very similar to that which causes kuru is
    (A) hepatitis.
    (B) HIV.
    (C) smallpox.
    (D) Creutzfeldt-Jacob.
    (E) chickenpox.

CHAPTER 17

Prokaryotes

337. If a bacterium is thermophilic, it means that it has a maximum growth rate at elevated temperatures, such as *Thermus aquaticus*, which grows best at around 72°C. Why would this organism NOT grow well at a mesophilic temperature of about 37°C?
    (A) Because lipids needed at elevated temperatures congeal at lower temperatures.
    (B) Because all thermophiles also require elevated salt levels.
    (C) Because thermophilic bacterial enzymes work best at the elevated temperatures.
    (D) Because a thermophile's cell wall freezes at mesophilic temperature ranges.
    (E) Because a thermophile never encounters such depressed temperatures, such a question is irrelevant.

338. What do the following three organisms have in common: *Staphylococcus aureus*, *Streptococcus pyogenes*, and *Neisseria gonorrhoeae*?
    (A) All are Gram positive.
    (B) All are obligate intracellular parasites.
    (C) All are cocci.
    (D) All are normal flora of the throat.
    (E) both A and B

339. The taxonomic system that has Archaebacteria as a major division is the
    (A) two-kingdom system.
    (B) three-kingdom system.
    (C) four-kingdom system.
    (D) five-kingdom system.
    (E) both A and B

340. Of the following, which is/are a known probable cause of meningitis in children?
    (A) *Neisseria meningitidis*
    (B) *Streptococcus bovis*
    (C) *Haemophilus influenzae*
    (D) both A and B
    (E) both A and C

341. An organism that has an axial filament and is noted for causing urinary and kidney infections is
    (A) *Ureaplasma urealyticum.*
    (B) *Staphylococcus aureus.*
    (C) *Rhizobium* sp.
    (D) *Leptospira* sp.
    (E) none of the above

342. The taxonomic level of tribe fits between
    (A) kingdom and division.
    (B) class and order.
    (C) genus and species.
    (D) order and family.
    (E) division and class.

343. Pathogenic bacteria are almost always
    (A) photosynthetic.
    (B) heterotrophic.
    (C) autotrophic.
    (D) methanogens.
    (E) both A and C

344. Which of the following characterize bacteria?
    (A) They all have cell membranes.
    (B) They all have nuclei.
    (C) They all have mitochondria.
    (D) Most have cell walls.
    (E) both A and D

345. DNA in bacteria is
    (A) circular.
    (B) linear.
    (C) single stranded.
    (D) both A and C
    (E) both B and C

346. The term *ubiquitous* means
    (A) very small.
    (B) only found underwater.
    (C) pathogenic.
    (D) can be found everywhere.
    (E) composed of bacteria.

347. The group of bacteria that is usually found under extreme conditions is normally classified as
    (A) eubacteria.
    (B) archaebacteria.
    (C) protests.
    (D) algae.
    (E) none of the above

348. Which of the following lacks a nucleus?
    (A) *Staphylococcus* sp.
    (B) *Plasmodium* sp.
    (C) *Giardia* sp.
    (D) *Paramecium* sp.
    (E) none of the above

349. The name of the organism that causes a sexually transmitted disease is
    (A) *Thermus aquaticus*.
    (B) *Entamoeba histolytica*.
    (C) *Streptococcus pyogenes*.
    (D) *Plasmodium* sp.
    (E) *Treponema pallidum*.

350. The simplest of the eukaryotes are the
    (A) protistans.
    (B) plants.
    (C) fungi.
    (D) animals.
    (E) bacteria.

351. Which of the following best describes a plasmid?
    (A) self-replicating
    (B) circular double-stranded DNA
    (C) lacking any structural gene segments
    (D) both A and B
    (E) none of the above

352. The method used by bacteria to replicate their DNA prior to cellular division is
    (A) alpha mode replication.
    (B) theta mode replication.
    (C) rolling circle replication.
    (D) ribosomal replication.
    (E) both A and C

353. The genetic map of *E. coli* was constructed using the phenotype(s)
    (A) Hfr$^+$.
    (B) F$'$.
    (C) F$^-$.
    (D) all of the above
    (E) both A and C

354. Which, if any, of the following describes transformation?
    (A) mRNA is transferred from cell to cell in a double-stranded form.
    (B) Protein is transferred from cell to cell in an uncoated form.
    (C) DNA is transferred in a bacteriophage.
    (D) RNA is transferred in a bacteriophage.
    (E) none of the above

355. Which of the following best describes a bacterial transposon?
    (A) 700–1,400 bp in length
    (B) 20–200 kbp in length
    (C) at both ends of coding genes
    (D) both B and C
    (E) none of the above

356. Transduction is the process whereby
    (A) DNA is transferred in a naked form.
    (B) protein is transferred in a naked form.
    (C) DNA is transferred in a bacteriophage.
    (D) RNA is transferred in a bacteriophage.
    (E) none of the above

357. Which of the following structures are NOT associated with bacteria?
    (A) pili
    (B) cilia
    (C) flagella
    (D) capsules
    (E) siderophores

CHAPTER 18

Protozoans

358. Four of the five organisms in the following list are classified as protozoans. Select the exception.
    (A) euglenoids
    (B) slime molds
    (C) bacteria
    (D) dinoflagellates
    (E) diatoms

359. The name of the organism that causes malaria is
    (A) *Treponema pallidum.*
    (B) *Entamoeba histolytica.*
    (C) *Streptococcus pyogenes.*
    (D) *Plasmodium* sp.
    (E) *Thermus aquaticus.*

360. Four of the five answer options are cellular. Select the exception.
    (A) parasitoid
    (B) diatom
    (C) trypanosome
    (D) ciliate
    (E) dinoflagellate

361. The name of the organism that causes amoebic dysentery is
    (A) *Treponema pallidum.*
    (B) *Entamoeba histolytica.*
    (C) *Streptococcus pyogenes.*
    (D) *Plasmodium* sp.
    (E) *Thermus aquaticus.*

362. An organism that lives in a single-celled form in the soil but congregates to form a macroscopic fruiting body is a(n)
    (A) fungus.
    (B) acellular slime mold.
    (C) protozoan.
    (D) cellular slime mold.
    (E) exceptionally large bacterium.

363. The vector that spreads *Trypanosoma* sp. is
    (A) the tsetse fly.
    (B) the mosquito.
    (C) fecal matter.
    (D) stagnant water.
    (E) none of the above

364. The form of life whose remains are used as dental abrasives in toothpaste is/are
    (A) diatoms.
    (B) dinoflagellates.
    (C) bacteria.
    (D) viruses.
    (E) chlorophytes.

365. Protozoans are dimorphic in that they
    (A) always undergo mitosis to form two daughter cells.
    (B) can exist in two forms: filaments and yeast.
    (C) can exist as either free-living or parasitic forms.
    (D) can exist as either amoeboid or ciliate forms.
    (E) none of the above

366. Some amoeboid protozoans live inside a glassy cage. This cage is called a
    (A) glacine.
    (B) cell wall.
    (C) plate.
    (D) kinetoplast.
    (E) test.

367. The white cliffs of Dover, England, are composed of
    (A) ciliates.
    (B) fossilized bacteria.
    (C) diatoms.
    (D) protozoan cysts.
    (E) lye produced by fungal digestion.

368. Pseudopodia are best associated with
    (A) slime molds.
    (B) amoeboid movement.
    (C) malaria.
    (D) tapeworms.
    (E) diarrhea.

369. Which of the following would best be used to treat an intestinal parasitic infection?
    (A) chlorine bleach
    (B) aspirin
    (C) miconazole
    (D) flagyl
    (E) penicillin

370. Which of the following structures are unique to protozoans?
    (A) kinetoplasts
    (B) nuclei
    (C) lysosomes
    (D) gametes
    (E) mycelia

371. The human protozoan disease that results in the formation of large disfiguring scars on the skin is
    (A) giardiasis.
    (B) leishmaniasis.
    (C) myiasis.
    (D) elephantiasis.
    (E) malaria.

372. The protozoan form best associated with survival under hostile conditions is called a
    (A) troph.
    (B) cyst.
    (C) test.
    (D) ciliate.
    (E) flagellate.

373. A form of reproduction unique to protozoans is
    (A) binary fission.
    (B) mitosis.
    (C) meiosis.
    (D) schizogony.
    (E) double fertilization.

374. In which form do the vast majority of protozoans exist?
    (A) free living
    (B) parasitic
    (C) symbiotic
    (D) mutualistic
    (E) pathogenic

375. Which form of metabolism do most protozoans utilize?
    (A) photosynthetic
    (B) lithotrophic
    (C) autotrophic
    (D) heterotrophic
    (E) both B and D

376. Which form of reproduction do protozoans lack?
    (A) mitosis
    (B) conjugation
    (C) asexual
    (D) merogony
    (E) meiosis

377. A protozoan disease spread by kissing bugs is called
    (A) malaria.
    (B) schistosomiasis.
    (C) Chagas disease.
    (D) giardiasis.
    (E) amoebiasis.

378. Which protozoan disease is best associated with water sources in which beavers make their homes?
    (A) amoebiasis
    (B) African sleeping sickness
    (C) giardiasis
    (D) elephantiasis
    (E) pinworm

CHAPTER 19

Fungi

379. A mushroom would be classified in the division
    (A) Ascomycota.
    (B) Basidiomycota.
    (C) Deuteromycota.
    (D) Oomycota.
    (E) Zygomycota.

380. Which fungus relies on extracellular digestion and absorption of energy-rich substances found in living organisms?
    (A) basidiomycetic
    (B) saprobic
    (C) parasitic
    (D) plasmodial
    (E) autotrophic

381. The type of organism that causes athlete's foot is a
    (A) fungus.
    (B) bacterium.
    (C) virus.
    (D) prion.
    (E) protozoan.

382. Which of the following would you expect to find in a grocery store?
    (A) *Agaricus bisporus*
    (B) *Saccharomyces cerevisiae*
    (C) dermatophytes
    (D) all of the above
    (E) both A and B

383. *Dimorphic* for fungi means
   (A) having two heads.
   (B) having two forms, yeast and filamentous.
   (C) changing from a bacteria to a protozoan.
   (D) changing from a protozoan to a yeast.
   (E) having two copies of every gene.

384. Another name for the Deuteromycota is
   (A) *Candida albicans.*
   (B) Fungi imperfecti.
   (C) deuteromycetes.
   (D) both B and C
   (E) both A and B

385. A fungus with a sexual stage consisting of a thickened spore would be classified in the division
   (A) Ascomycota.
   (B) Basidiomycota.
   (C) Deuteromycota.
   (D) Oomycota.
   (E) Zygomycota.

386. A fungus with sexual stage spores within a sac would be classified in the division
   (A) Ascomycota.
   (B) Basidiomycota.
   (C) Deuteromycota.
   (D) Oomycota.
   (E) Zygomycota.

387. Which of the following is NOT an asexual fungal spore form?
   (A) arthrospores
   (B) chlamydospores
   (C) blastospores
   (D) conidiospores
   (E) none of the above

388. Which could be used to differentiate a fungus from a multicellular alga?
    (A) the presence or absence of chloroplasts
    (B) the presence or absence of mitochondria
    (C) the composition of the cell wall
    (D) all of the above
    (E) both A and C

389. Fungi are associated with the production of all of the following foods except
    (A) beer.
    (B) cheese.
    (C) sauerkraut.
    (D) coffee.
    (E) bread.

390. The material used by fungi to manufacture their cell wall is
    (A) chitin.
    (B) cellulose.
    (C) peptidoglycan.
    (D) silica.
    (E) calcium carbonate.

391. Lichens are a form of life that are composed of both algal and fungal cells. The relationship between these cells is described as
    (A) parasitic.
    (B) symbiotic.
    (C) saprobic.
    (D) cytophagic.
    (E) autotrophic.

392. Yeasts reproduce by a process called
    (A) schizogony.
    (B) budding.
    (C) binary fission.
    (D) merogony.
    (E) meiosis.

393. What is the relationship between hyphae and mycelia?
   (A) Mycelia consist of more than one cell, while hyphae have only one.
   (B) Mycelia are related to fungi, while hyphae are related to algae.
   (C) Mycelia are composed of hyphae.
   (D) Mycelia contain chitin; hyphae contain cellulose.
   (E) Mycelia and hyphae are unrelated.

394. If fungi did not exist, what geologic cycle would be greatly disrupted?
   (A) carbon cycle
   (B) nitrogen cycle
   (C) oxygen cycle
   (D) water cycle
   (E) none of the above

395. What is the unique characteristic of oomycetes?
   (A) They appear to be a transition group between viruses and fungi.
   (B) They are genetically unrelated to fungi but have long been classified with them.
   (C) While classified as fungi, their cell walls are composed of silica.
   (D) They are the only fungi that produce motile spores.
   (E) Their mycelia form a characteristic square pattern.

396. The Death Cap is a toxic fungus readily identified as a
   (A) mold.
   (B) mushroom.
   (C) morel.
   (D) puffball.
   (E) mildew.

397. Pigs are used by European farmers to find the tree-root-associated delicacy known as the
   (A) truffle.
   (B) morel.
   (C) table mushroom.
   (D) porcini mushroom.
   (E) shiitake mushroom.

398. The structure we identify as a mushroom is actually best associated with
    (A) energy storage.
    (B) light gathering.
    (C) shade production.
    (D) water storage.
    (E) reproduction.

399. Which form of reproduction is best associated with the Fungi imperfecti?
    (A) meiosis
    (B) binary fission
    (C) mitosis
    (D) schizogony
    (E) conjugation

CHAPTER **20**

Plants

400. Plant gametophytes

    (A) are haploid.
    (B) are diploid.
    (C) form gametangia.
    (D) both A and C
    (E) both B and C

401. _____ produce a seta based at a foot.

    (A) Ferns
    (B) Mosses
    (C) Angiosperms
    (D) Gymnosperms
    (E) both C and D

402. Monocots have

    (A) one cotyledon.
    (B) netted veins in their leaves.
    (C) endosperm in their seeds.
    (D) all of the above
    (E) A and C only

403. _____ produce rhizomes and fronds.

    (A) Ferns
    (B) Mosses
    (C) Angiosperms
    (D) Gymnosperms
    (E) both C and D

404. Herbaceous growth with flowers of multiples of three petals describes
   (A) monocots.
   (B) dicots.
   (C) ferns.
   (D) conifers.
   (E) none of the above

405. A double fertilization process describes
   (A) ferns.
   (B) mosses.
   (C) angiosperms.
   (D) gymnosperms.
   (E) both C and D

406. _____ produce a prothallus.
   (A) Ferns
   (B) Mosses
   (C) Angiosperms
   (D) Gymnosperms
   (E) both C and D

407. Which, if any, of the following is NOT a part of a flower?
   (A) anther
   (B) sori
   (C) style
   (D) ovary
   (E) All of the above are parts of a flower.

408. _____ produce seeds.
   (A) Ferns
   (B) Mosses
   (C) Angiosperms
   (D) Gymnosperms
   (E) both C and D

409. *Alternating generations* is a term used to describe
   (A) fungi.
   (B) protistans.
   (C) algae.
   (D) bacteria.
   (E) none of the above

410. _____ produce fruit.
    (A) Ferns
    (B) Mosses
    (C) Angiosperms
    (D) Gymnosperms
    (E) both C and D

411. "Multicellular photosynthetic autotrophs" describes
    (A) monocots.
    (B) dicots.
    (C) ferns.
    (D) conifers.
    (E) all of the above

412. A lichen is a symbiotic relationship between a fungus and
    (A) another fungus.
    (B) a plant.
    (C) bacteria.
    (D) algae.
    (E) none of the above

413. The tissue that protects the root apical meristem is the
    (A) bud primordial.
    (B) vascular cambium.
    (C) leaf primordia.
    (D) cork cambium.
    (E) none of the above

414. The plant vascular tissue cells that have flat ends are the
    (A) collenchyma.
    (B) tracheids.
    (C) vessel elements.
    (D) sclerenchyma.
    (E) none of the above

415. Leaf stomata are best associated with
    (A) transpiration.
    (B) food production.
    (C) protection from predators.
    (D) bud formation.
    (E) lateral stem growth.

416. In plants, oxygen is
  (A) consumed within mitochondria.
  (B) released by chloroplasts.
  (C) detoxified by peroxisomes.
  (D) released in the form of ozone.
  (E) both A and B

417. The expansion of bark in tree trunks is caused by
  (A) transpiration.
  (B) the movement of sap within the phloem.
  (C) lateral meristems.
  (D) bud primordial.
  (E) the absorption of water from the atmosphere.

418. Which organelle or structure is NOT associated with plants?
  (A) lysosomes
  (B) mitochondria
  (C) vacuoles
  (D) chloroplasts
  (E) endosomes

419. Some of the oldest living individual organisms on the planet include plants classified as
  (A) gymnosperms.
  (B) bryophytes.
  (C) hornworts.
  (D) monocots.
  (E) angiosperms.

420. What becomes of the eight nuclei produced by an angiosperm megaspore mother cell?
  (A) One becomes an egg, five become embryonic sacs, and two degenerate.
  (B) One becomes an egg, two become polar nuclei, and five degenerate.
  (C) One becomes an embryonic sac, two become polar nuclei, and five degenerate.
  (D) Two degenerate, two become polar nuclei, two become eggs, and two become embryonic sacs.
  (E) One degenerates, two become eggs, and five develop into embryonic sacs.

CHAPTER 21

Animals

421. Which of the following is the most inclusive of the group?
    (A) cells
    (B) tissues
    (C) organs
    (D) glands
    (E) organ systems

422. Which germ layer produces the external covering of the body?
    (A) ectoderm
    (B) mesoderm
    (C) neural crest
    (D) blastosphere
    (E) endoderm

423. The right and left halves of the body are divided by the _____ plane.
    (A) frontal
    (B) midsaggital
    (C) transverse
    (D) abdominal
    (E) saggital

424. _____ are the simplest animals that have nerve cells.
    (A) Chordates
    (B) Cnidarians
    (C) Platyhelminthes
    (D) Poriferans
    (E) Annelids

425. The most common intermediate hosts for flukes are
  (A) annelids.
  (B) insects.
  (C) snails.
  (D) toads.
  (E) squirrels.

426. A true coelom is found in which of the following?
  (A) arthropods
  (B) protozoans
  (C) nemertineans
  (D) molluscs
  (E) platyhelminths

427. Bivalves are
  (A) protozoans.
  (B) parasites.
  (C) diatoms.
  (D) filter feeders.
  (E) fish.

428. _____ is the material that forms the covering of tunicates.
  (A) Chitin
  (B) Cellulose
  (C) Bone
  (D) Peptidoglycan
  (E) Protein

429. The _____ period is when labyrinthodonts are thought to have first evolved into reptiles.
  (A) Permian
  (B) Precambrian
  (C) Ordovician
  (D) Devonian
  (E) Carboniferous

430. The most important purpose of the scales of a reptile is to
  (A) produce secretions.
  (B) aid in reproduction.
  (C) prevent desiccation.
  (D) prevent cannibalism.
  (E) assist in hatching.

431. _____ is one characteristic not possessed by all mammals.
- (A) Lactation
- (B) Hair
- (C) Teeth that are differentiated
- (D) A muscular diaphragm
- (E) A placenta

432. Marsupials are mostly found in
- (A) Australia.
- (B) New Guinea.
- (C) Africa.
- (D) Asia.
- (E) Hawaii.

433. Mammals are thought to have evolved from the group of reptiles called
- (A) therapsids.
- (B) thecodonts.
- (C) pterodactyls.
- (D) crocodilians.
- (E) cotylosaurs.

434. When comparing primates with other animals, which sense is less effective in the former?
- (A) touch
- (B) smell
- (C) sight
- (D) hearing
- (E) taste

435. The _____ is known for its behavior of laying its eggs during the highest point of the highest tide.
- (A) shark
- (B) grunion
- (C) sea turtle
- (D) lamprey eel
- (E) clownfish

436. Circadian rhythm appears to be most closely linked to the function of the _____ gland.
- (A) adrenal
- (B) thyroid
- (C) parathyroid
- (D) hypothalamus
- (E) pineal

437. The sense of _____ is the imprinting key used by sheep.
- (A) sight
- (B) taste
- (C) touch
- (D) smell
- (E) hearing

438. Insight learning is a characteristic best exhibited by
- (A) dogs.
- (B) birds.
- (C) primates.
- (D) horses.
- (E) water buffalo.

439. When an animal adheres to a lunar cycle, it is most likely to be one that is
- (A) marine.
- (B) terrestrial.
- (C) subterranean.
- (D) parasitic.
- (E) capable of flight.

440. Nematocysts are involved in
- (A) circulation.
- (B) reproduction.
- (C) predation.
- (D) homeostasis.
- (E) photosynthesis.

CHAPTER 22

Ecological Principles

441. _____ is the study of the interactions of organisms within their environment.
- (A) Community
- (B) Ecosystem
- (C) Population
- (D) Ecology
- (E) none of the above

442. A(n) _____ is all species within a habitat.
- (A) community
- (B) ecosystem
- (C) population
- (D) ecology
- (E) none of the above

443. A(n) _____ is a group of the same species living in a specific habitat.
- (A) community
- (B) ecosystem
- (C) population
- (D) ecology
- (E) none of the above

444. A(n) _____ is a community in relationship to its environment.
- (A) community
- (B) ecosystem
- (C) population
- (D) ecology
- (E) none of the above

445. Which of the following is an omnivore?
 (A) carrot
 (B) finch
 (C) cat
 (D) bear
 (E) osprey

446. Which biome exhibits the greatest amount of species diversity?
 (A) tundra
 (B) tropical rain forest
 (C) desert
 (D) temperate deciduous forest
 (E) grassland

447. Which of the following is NOT true concerning an ecosystem?
 (A) Energy flow is dependent on the number of producers relative to consumers.
 (B) The more efficient the producers, the greater the energy flow.
 (C) Smaller ecosystems are more stable than large ones.
 (D) Ecosystems change with age.
 (E) Flexibility is a characteristic more common to large ecosystems.

448. When two species are totally dependent upon each other for survival, the relationship is known as
 (A) cooperation.
 (B) parasitism.
 (C) interspecies competition.
 (D) symbiosis.
 (E) commensalism.

449. _____ is when one species provides another species benefits to its own detriment.
 (A) Parasitism
 (B) Commensalism
 (C) Symbiosis
 (D) Interspecies competition
 (E) Exploitive competition

450. What term best describes the relationship flies have with humans?
   (A) obligate
   (B) parasitic
   (C) mutualistic
   (D) free living
   (E) commensal

451. A relationship between two species where both are harmed is called
   (A) mutualism.
   (B) symbiosis.
   (C) commensalism.
   (D) neutral interaction.
   (E) competition.

452. A(n) _____ is the most favorable habitat for an organism.
   (A) biome
   (B) ecosystem
   (C) environment
   (D) niche
   (E) stratum

453. An insect that deposits its eggs on another so its larvae can consume it is called a
   (A) parasite.
   (B) parasitoid.
   (C) host.
   (D) degenerate.
   (E) competitor.

454. A good example of a species that acts as a social parasite is the
   (A) seagull.
   (B) heron.
   (C) cowbird.
   (D) hummingbird.
   (E) crow.

455. Which is NOT a factor that limits growth?
   (A) mutualism
   (B) parasitism
   (C) emigration
   (D) resource partitioning
   (E) predation

456. If an organism is found in an area that appears to have no predator, then it is probably
    (A) at trophic level 3.
    (B) a predator itself.
    (C) a very effective producer.
    (D) an invasive species.
    (E) a parasite.

457. When one community follows another in a predictable order, the process is called
    (A) resource partitioning.
    (B) competition.
    (C) community succession.
    (D) ecologic success.
    (E) diversification.

458. When an ecosystem reaches a self-sustaining level of growth, then
    (A) the entire system has become dependent on one very productive species.
    (B) it has become a climax community.
    (C) it exhibits a J-shaped curve.
    (D) all species have become mutualistic.
    (E) all predator species have been eliminated.

459. The probability of an ecosystem developing where each species is symbiotic with all others within the system is
    (A) zero.
    (B) 100% given enough time.
    (C) low assuming normal mutation rates.
    (D) high given that all the species are bacteria.
    (E) 100% given enough species.

460. If an immigrant species is successful within a new ecosystem, then it
    (A) has become a predator.
    (B) must be a decomposer.
    (C) must be crepuscular.
    (D) most likely mimics a species already present.
    (E) has become adapted.

CHAPTER 23

Population Ecology

461. Terms used to describe the distribution of populations include
   (A) clumped.
   (B) random.
   (C) uniform.
   (D) all of the above
   (E) both A and B

462. _____ can increase population size.
   (A) Emigration
   (B) Immigration
   (C) Biotic potential
   (D) both A and C
   (E) both B and C

463. A _____ survivorship curve is used to describe large terrestrial animals.
   (A) Type I
   (B) Type II
   (C) Type III
   (D) Type IV
   (E) either C or D

464. The human population on the earth reached _____ around 1930.
   (A) one billion
   (B) two billion
   (C) four billion
   (D) six billion
   (E) ten billion

465. An "S-shaped" growth curve describes
- (A) exponential growth.
- (B) logistic growth.
- (C) suicidal growth.
- (D) zero growth.
- (E) none of the above

466. _____ is a growth factor that is density independent.
- (A) Predation
- (B) Disease
- (C) Floods
- (D) Parasitism
- (E) All of the above are density independent.

467. _____ can decrease population size.
- (A) Emigration
- (B) Immigration
- (C) Biotic potential
- (D) both A and C
- (E) both B and C

468. A _____ survivorship curve is used to describe organisms that produce huge numbers of offspring with little to no nurturing.
- (A) Type I
- (B) Type II
- (C) Type III
- (D) Type IV
- (E) either C or D

469. The human population on the earth is now closest to
- (A) one billion.
- (B) two billion.
- (C) four billion.
- (D) seven billion.
- (E) ten billion.

470. A "J-shaped" growth curve describes
- (A) exponential growth.
- (B) logistic growth.
- (C) suicidal growth.
- (D) zero growth.
- (E) none of the above

471. Humans add to the population by about _____ people per hour.
    (A) 100
    (B) 1,000
    (C) 10,000
    (D) 100,000
    (E) 1,000,000

472. A _____ survivorship curve is used to describe small terrestrial animals.
    (A) Type I
    (B) Type II
    (C) Type III
    (D) Type IV
    (E) either C or D

473. _____ is considered a density-dependent factor regulating population growth.
    (A) Drought
    (B) Flood
    (C) Competition
    (D) Global cooling
    (E) Wildfire

474. The most intense form of competition is identified as
    (A) predation.
    (B) interspecies.
    (C) parasitism.
    (D) intraspecies.
    (E) symbiosis.

475. Which of the following is NOT a reason for rapid population growth?
    (A) earlier reproductive age
    (B) reduction of limiting factors
    (C) emigration
    (D) moving into new habitats
    (E) immigration

476. A group of individuals of the same species and the same age defines a
    (A) cohort.
    (B) population.
    (C) cluster.
    (D) mating group.
    (E) survivorship curve.

477. The greatest cause of human deaths is
    (A) cancer.
    (B) war and battle.
    (C) accidents.
    (D) heart disease.
    (E) starvation and malnutrition.

478. _____ is the most common population distribution pattern.
    (A) Clumped
    (B) Random
    (C) Uniform
    (D) Stratified
    (E) Curved

479. Population growth is most commonly expressed by
    (A) population density.
    (B) doubling time.
    (C) biomass.
    (D) carrying capacity.
    (E) immigration.

480. What happens when a population exceeds the carrying capacity?
    (A) Immigration increases.
    (B) Biotic potential increases.
    (C) Exponential death occurs.
    (D) A new trophic level becomes available.
    (E) The population begins exponential growth.

CHAPTER 24

Communities and Ecosystems

481. The only one of the following that is NOT considered a member of an ecosystem is the
    (A) consumer.
    (B) producer.
    (C) environment.
    (D) decomposer.
    (E) none of the above

482. _____ does not cause an increase of carbon dioxide within the atmosphere.
    (A) Combustion
    (B) Methane-generated power
    (C) Human respiration
    (D) Volcanoes
    (E) Photosynthesis

483. The only carbon available for photosynthesis is found in
    (A) the atmosphere.
    (B) carbonates dissolved within the ocean.
    (C) vegetation.
    (D) organic molecules.
    (E) fossil fuels.

484. Which of the following is NOT considered part of the nitrogen cycle?
    (A) nitrate
    (B) ammonia
    (C) nitrogen gas
    (D) nitrous oxide
    (E) nitrite

485. Legumes are important components of the nitrogen cycle because they
  (A) decay into nitrogen-rich soil.
  (B) harbor nitrogen-fixing bacteria within their root nodules.
  (C) release nitrogen gas during respiration.
  (D) serve as a primary food source for methane-releasing cattle.
  (E) release nitrates into the atmosphere.

486. _____ -containing compounds tend to remain deposited on the ocean floor because they are relatively insoluble.
  (A) Oxygen
  (B) Carbon
  (C) Nitrogen
  (D) Hydrogen
  (E) Phosphorus

487. When organic materials decay the primary nitrogen by-product is (are)
  (A) ammonia.
  (B) amino acids.
  (C) nitrogen gas.
  (D) nitrates.
  (E) nitrites.

488. Herbivores are considered to reside within the _____ trophic level.
  (A) bottom
  (B) second
  (C) first
  (D) top
  (E) third

489. The greatest amount of biomass would normally be expected to be found within the _____ trophic level.
  (A) fifth
  (B) second
  (C) first
  (D) fourth
  (E) third

490. A _____ is typically used to represent the distribution of trophic level biomass within an ecosystem.
    (A) circle
    (B) diamond
    (C) sphere
    (D) pyramid
    (E) trapezoid

491. The greatest limitation on energy flow within an ecosystem is (are)
    (A) density-dependent factors.
    (B) the availability of producers to manufacture carbon-containing compounds.
    (C) the mean temperature.
    (D) density-independent factors.
    (E) the relative numbers of predators, parasites, and prey.

492. The planetary reservoir of nitrogen is
    (A) the atmosphere.
    (B) terrestrial organisms.
    (C) the oceans.
    (D) ammonia.
    (E) decaying vegetation.

493. About what percentage of the energy content contained within the biomass is passed from one trophic level to the next?
    (A) 50%
    (B) 2%
    (C) 25%
    (D) 10%
    (E) 15%

494. Most terrestrial communities will NOT exceed _____ trophic levels.
    (A) one
    (B) three
    (C) two
    (D) four
    (E) six

495. The primary reservoir of carbon within the ocean is found in the form of
    (A) carbon dioxide.
    (B) calcium carbonate.
    (C) DNA.
    (D) proteins.
    (E) magma.

496. The least amount of water on the earth is contained within
    (A) aquifers.
    (B) glaciers.
    (C) saltwater.
    (D) freshwater.
    (E) groundwater.

497. Organisms that are chemosynthetic contribute to an ecosystem by being
    (A) primary producers.
    (B) tertiary consumers.
    (C) decomposers.
    (D) nitrogen fixers.
    (E) secondary producers.

498. Carbon dioxide is present in the atmosphere at about
    (A) 17 parts per billion.
    (B) 35 parts per thousand.
    (C) 170 parts per billion.
    (D) 350 parts per million.
    (E) 3.5 parts per million.

499. _____ is the organic process that produces N_2 gas.
    (A) Nitrogen fixation
    (B) Ammonification
    (C) Denitrification
    (D) Decomposition
    (E) Nitrification

500. What is biologic magnification?
    (A) the buildup of carbonate deposits on the ocean floor
    (B) the process whereby toxins increasingly accumulate at higher levels in the food chain
    (C) the ever-increasing importance of one species as ecosystems become less diverse
    (D) the buildup of carbon dioxide that causes global warming
    (E) the spread of a parasite within a population as the density reaches the carrying capacity

ANSWERS

Chapter 1: Cell Chemistry

1. (B) Fatty acids are the monomers that are used to construct lipids. Monosaccharides, or simple sugars, are the monomers used to construct polysaccharides. Nucleotides are the monomers used to construct nucleic acids such as DNA and RNA. Vitamins are normally cofactors required for the functioning of some enzymes. Proteins, also known as polypeptides, are polymers of amino acids.

2. (D) Monosaccharides are connected through bonds that connect the number 1 carbon atom of one sugar with the number 4 carbon atom of the next sugar, a connection identified as a 1,4 glycoside linkage. However, two different bond configurations are possible, either an α linkage or a β linkage. The enzymes possessed by most animals can break only the α-1,4 linkage found in starch but are incapable of breaking the β-1,4 linkage of cellulose.

3. (C) The pH scale is a logarithmic measurement of the number of hydrogen ions (H^+) in a water-based solution. This means that a tenfold increase of H^+ will decrease the pH from 5.0 to 4.0. To decrease the pH by another full factor to 3.0 would require the addition of 10 times the initial amount added; here an additional 100 drops would be required, not just 10. The addition of 10 drops of the original concentration would drop the pH by only approximately 10%.

4. (E) Proteins can act as enzymes and are involved in cellular motion, structure, and membrane transport. They can also be used for energy storage and control of gene expression. Even though histone proteins are associated with the organization of eukaryotic DNA, proteins are not associated with the storage of genetic information, as this is characteristic of nucleic acids.

5. (A) DNA is an organic molecule that is always double stranded with the exception of a few viral genomes. The structure is maintained by complementary base pairing and is replicated in a semiconservative manner, meaning that each cell following cell division inherits one original strand and one newly synthesized complementary copy. However, when describing DNA, the structure is identified as being antiparallel, as the molecular orientation of each strand is in the opposite direction of its partner, with one going 3'-5' and the other 5'-3'.

6. (B) Carbohydrates and lipids contain only carbon, hydrogen, and oxygen in their chemical structure. In addition to these three elements, nucleic acids require nitrogen as well. Proteins are constructed from amino acids, and two of these essential amino acids, methionine and cysteine, contain sulfur. Cysteine is associated with disulfide crosslinking commonly essential for maintaining proper protein structure and function. Cell walls are composed of carbohydrates and thus do not need sulfur.

7. (D) Water molecules are strongly attracted to each other and thus tend to cluster into droplets. This tremendous cohesive ability is not due to water's small mass or size as similarly small hydrogen, nitrogen, carbon dioxide, and oxygen molecules lack this characteristic. The cohesive ability of water is due to its partial molecular polarity, where the shared electrons tend to congregate on the oxygen side of the molecule, making it partially negatively charged and making the hydrogen side partially negatively charged. Water's freezing point is dictated by this cohesive ability, not the other way around.

8. (E) *Amphipathic* refers to a molecule that is uncharged on one end and charged on the other end. The nonpolar end is usually lipid based and hydrophobic. Because the other end is charged, it interacts well with the water molecules that commonly surround it and is thus hydrophilic. This term is commonly used to describe the various phospholipids that comprise the cell membrane.

9. (C) When a protein is synthesized in a ribosome, one amino acid after another is added to the emerging polypeptide chain. This amino acid sequence is referred to as the primary protein structure. Commonly the interactions of the amino acid side chains will cause the emerging chain to form a tight spiral structure or a repetitive folding pattern, referred to as an α-helix or β-pleating, classified as a secondary structure. When the protein is released, it folds into its three-dimensional tertiary structure. Often multiple polypeptide chains are assembled and crosslinked for stability in a form referred to as its quaternary structure.

10. (A) Distilled water has a pH of 7.0 with an equal balance of hydrogen (H^+) and hydroxyl (OH^-) ions. Acidic solutions have an excess of H^+ ions, ranging from battery acid at about pH = 1.0 to milk at about pH = 6.5. Both vinegar and carbonated soft drinks are toward the middle of this range and average about pH = 3.0. Alkaline solutions, on the other hand, have excess OH^- ions, ranging from blood at pH = 7.4 to bleach at pH = 9.0 to oven cleaner at pH = 13.0 and lye at pH = 14.0.

11. (D) A sugar is simply any carbohydrate that tastes sweet. Commonly, these are mono- or disaccharides and either pentoses or hexoses. Disaccharides, or two simple sugars connected by a glycoside bond, include sucrose, lactose, and maltose. Pentoses, simple sugars with five-carbon atoms, include ribose, deoxyribose, xylose, and arabinose. Hexoses, simple sugars with six-carbon atoms, include glucose (also known as dextrose), fructose, mannose, and galactose.

12. (C) A hydrophobic molecule is one that is nonpolar and tends to exclude water. The term literally means "water fearing." Sodium chloride readily disassociates in water, as the ionic bonds holding the molecule together are overcome by the partial polarity of the surrounding water. CO_2 remains intact, but as a gas, dissolves in water as well. Both proteins and sucrose are hydrophilic and readily dissolve in water.

13. (E) Halogens include the elements fluorine, chlorine, bromine, iodine, and astatine. These are located next to the inert gases on the right side of the periodic table. The primary characteristic that all of these elements have in common is that they are one electron short of filling their outer orbital and thus strongly attract that missing electron from adjacent atoms. Any substance that loses an electron to chlorine is oxidized, and when this occurs to a protein, a significant change occurs in its tertiary structure.

14. (B) Glycogen is a branched polymer of glucose produced in the liver of higher animals. Starch is a branched polymer of glucose formed in plants, while amylose is the equivalent unbranched form. All three are energy storage products.

15. (A) Buffers are used to maintain a fairly constant pH in a solution containing biologic materials. This is because the biologic functions of those materials are greatly affected by the balance of hydrogen and hydroxyl ions, particularly the functions of proteins. Buffers have the ability to remove these free ions from the solution up to a point of saturation, at which time any additions of acid or base will change the overall pH of the solution.

16. (A) Bile salts are manufactured in the liver from a cholesterol precursor and then stored in the gallbladder. These salts are a major component of the bile and assist in the breakdown and absorption of lipids in the small intestine. Reproductive hormones, however, have no association with the digestive process or the liver. All three do share a common four-carbon ring structure that defines a steroid.

17. (C) Bacterial cell walls are composed of peptidoglycan made up of alternating N-acetylmuramic acid (NAM) and N-acetylglucosamine (NAG) residues crosslinked with peptide chains. The basic component of both the fungal cell wall and the insect exoskeleton is chitin, which is a polymer of only NAG. However, NAM and NAG are both polysaccharides.

18. (B) Some lipids contain nitrogen or phosphorus but not all. The same is true for lipids linked to amino acids. The bonds that link the fatty acids (which vary from one chain in a monoglyceride, to two in a diglyceride, to three in a triglyceride) to a common glycerol core are an ester linkage, not a glycoside linkage.

19. (E) An amino acid is an organic compound with a hydrogen atom, a carboxyl group (COOH), an amino group (NH_2), and any of a variety of organic side groups (or hydrogen atoms as in the case of glycine) appended to a central carbon atom. Amino acids may be metabolized into ketone bodies but are not classified as ketones, which are organic molecules with a carbonyl group (C=O) bound to two other carbon groups.

20. (D) As a halogen, iodine is a strong oxidizer that serves as a disinfectant when at high concentrations by denaturing proteins, thus rendering them ineffective for their normal function. Nutritional iodine is required at low dietary levels in order for mammals to synthesize the thyroid hormones T_3 and T_4, which are essential for the regulation of cellular metabolism.

21. (C) Carbon atoms will always form four total bonds with other atoms. In some cases they are capable of forming double or triple bonds, which count as one or two additional bonds per carbon. Thus in $H_2C=CH_2$, both carbons are connected to two separate hydrogen atoms and to the adjacent carbon atom twice, for a total of four bonds each.

Chapter 2: Cell Structure and Function

22. (D) All cells have proteins embedded in their membranes, and although the specific proteins might vary between the two groups, they also have a lot in common. While the composition of the cell walls of both groups vary, the presence of the cell wall does not differentiate them. Both groups can metabolize some toxic materials to some extent or another.

While fungi always lack photosynthetic pigments, only a few bacterial groups possess them. When it comes to the presence of a nucleus, all fungi are eukaryotic, while all bacteria are prokaryotic.

23. **(B)** A cell membrane is composed of two leaflets of phospholipids, also known as a phospholipid bilayer. However, when visualized by electron microscopy, three layers are commonly visualized: the darker hydrophilic heads that line the two external layers and the lighter stained central hydrophobic region. Thus the cell membrane is trilaminar.

24. **(C)** Neither group produces estrogen or is anaerobic. While many fungi and insects are associated with the soil, both have numerous examples of genera that are not. Fungi are dimorphic, having both single cellular yeast and multicellular hyphal forms, but the term is not applied to insects. The cells walls of all fungi and the hard exoskeleton of insects are composed of the polysaccharide chitin.

25. **(A)** Both symport and antiport are mechanisms used by cells to transport substances through a plasma membrane utilizing membrane-bound proteins. Osmosis is the flow of water molecules through a membrane. Vesicular transport is utilized by eukaryotic cells to move substances from the Golgi apparatus to other areas of the cell following a microtubule framework. Substances entering or leaving the nucleus must be escorted through nuclear pores by specific escort proteins.

26. **(D)** The term *continuous* refers to a connected structure. The Golgi, mitochondria, and cell membrane are completely independent from the endoplasmic reticulum (ER). Ribosomes, responsible for protein synthesis, may associate with the ER when these proteins will be either secreted from the cell or membrane bound, but the ribosomes also can function while floating free in the cytosol. The membrane structure of the ER is always connected to the outer membrane of the nucleus.

27. **(E)** Cell walls are found surrounding all fungal and plant cells and around most bacterial cells, but are absent in animal and protozoan cells. Nuclei and mitochondria are found in all eukaryotic cells but are missing in bacteria. A cytoskeleton is found in all eukaryotes and in some, but not most, bacteria. Since all cells manufacture proteins, all cells must have ribosomes.

28. **(D)** Proteins are manufactured by ribosomes in the cytosol. Some of these ribosomes will migrate to the ER for assembly within its lumen. When synthesis is complete, some modifications to the proteins may take place within the ER where most will then be moved to the Golgi for additional modifications. Eventually the Golgi will send these proteins off by vesicular transport to other places within the cell, making it like a factory loading dock.

29. **(C)** All eukaryotic cells have nuclear DNA organized as chromatin, trilaminar cell membranes, and mitochondria. While plant cells do have cell walls, they are composed of cellulose, not chitin. What distinguishes plant cells from all others is the presence of photosynthesizing plastids called chloroplasts.

30. (A) Chromosomes carry the genes that transmit hereditary traits. Endosomes are transport vesicles moving materials around within the cell. Ribosomes are structures composed of proteins and RNA used for protein synthesis. While lysosomes contain many degradative materials, their function is to destroy phagocytosed cells, not toxins.

31. (B) Proteins typically fall in size from 5–10 nm. The vast majority of viruses fall between 20–200 nm. Most mitochondria are bacterial in size, falling between 1–2 μm. The nucleus is four to five times the size of mitochondria, or between 8–10 μm. The largest of those listed, the ovum, is about 1 mm in size.

32. (C) Amoebas are protozoans and do not have cellular structures associated with animal muscle cells, such as myofibrils and sarcolemmas. Microtubules are associated with the movement of flagella, cilia, and organelles. Intermediate filaments are associated with maintenance of cell shape. Microfilaments are associated with changes in cell shape.

33. (B) Mitochondria and chloroplasts are thought to have originated as bacteria that began a symbiotic relationship with a primordial ancestral cell. This endosymbiotic theory is supported by the fact that they are bacterial in size and contain basic energy-producing metabolic pathways. Additionally, they have their own bacteria-sized ribosomes.

34. (D) In biology the indicated letter "S" following a number usually refers to the svedberg unit, which is a laboratory measure of molecular density. The ribosomes of bacteria and eukaryotes are identical in function, similar in structure, but different in size. The bacterial ribosomes are composed of a large subunit (50S) and a small subunit (30S); when combined, the intact bacterial ribosome is 70S. The respective identifiers for the eukaryotic ribosome are 60S, 40S, and 80S.

35. (E) The structure of the bacterial flagellum and pilus is very simple when compared to the eukaryotic flagellum and cilium. Both of these latter structures are composed of microtubules arranged in a consistent pattern of nine microtubules surrounding a pair of microtubules in a circular pattern.

36. (C) The ribosome, regardless of whether it is found in eukaryotic cells in its 80S form or in prokaryotes in its 70S form, is always associated with the translation of the sequences found in mRNA strands into protein sequences. The ribosomal subunits are composed of one to three rRNA strands surrounded by between 29 and 41 ribosomal proteins.

37. (C) The term *Taq* is generally used to identify the DNA polymerase enzyme produced by a thermophilic archaebacterium. Topoisomerase is an enzyme required to stabilize the unwinding DNA helix during DNA replication. Both α- and β-subunits are required in order for the holoenzyme to have complete function. However, it is the σ-subunit that is first required to signal the proper location for the holoenzyme assembly.

38. (B) Only about 2% of the human genome contains codes for the approximately 20,000 to 30,000 genes required to maintain all cellular functions in cells. The remaining 98% includes noncoding sequences such as pseudogenes and introns, and mobile genetic elements such as transposons and retrotransposons. All in all, it would measure about 1.8 meters in length.

39. (E) Answer C is incorrect; the 7-meG cap that is required is added to the 5′ end of the molecule, not the 3′ end where the poly-A tail is attached. This makes answer D incorrect also. Answer E is the best choice because it is the most complete answer of those provided.

40. (A) *Taq* is used to identify a gene product from *Thermus aquaticus*, usually its DNA polymerase enzyme associated with use in the polymerase chain reaction DNA amplification technique. Similarly, *Eco*RI identifies a specific restriction endonuclease used in genetic engineering. The *glu* operon is constitutive, as it is always in use in a cell, and the *trp* operon is the best example of an operon that is repressible. The *lac* operon is induced when a bacterial cell is starving and has only lactose available as a carbon source.

41. (C) Oncogenes are associated with cancer, which is produced by any cell that loses its ability to regulate its own growth. A proto-oncogene is therefore any growth regulatory gene that experiences a mutation that either causes the resulting gene product to be ineffective in growth regulation or that interferes with the control on growth managed by another protein.

42. (E) DNA is found in the nuclei of all eukaryotic cells, in the nuclear region of prokaryotic cells, and as the genome in some viruses. It also can be found within nuclear regions of both chloroplasts and mitochondria, but never in ribosomes, which are composed of just proteins and rRNA. This makes answer E correct.

Chapter 3: Cell Membrane and Function

43. (D) Phospholipids have a hydrophilic head that is attracted to water and a hydrophobic tail that is repelled by water. When in an aqueous solution, these hydrophobic tails will cluster together as far from the water molecules as possible, which permits the hydrophilic heads to freely associate with the surrounding water. This produces a roughly spherical structure called a micelle.

44. (C) A cell will expend energy only when the expense offers a significant advantage to the cell; otherwise it will try to utilize more passive transport mechanisms, which include coupled transport. Both exocytosis and endocytosis require energy expense in order to manipulate membrane shape. Pumping substances across (or through) a membrane requires energy expense as well, which includes the sodium-potassium pump and the pumping of hydrogen ions against a gradient.

45. (D) Cell membranes are constructed as a basic phospholipid bilayer arranged as a fluid mosaic. However, many variations within the specific phospholipids exist and other components are also frequently inserted as necessary. These include numerous variations in proteins, glycoproteins, and glycolipids as well as cholesterol in the case of animals.

46. (A) The cell membrane serves as the controller of cellular communications and interaction with the world outside. Large materials and molecules cannot pass through the membrane unaided, and the membrane is impermeable to ions of any size. Water and molecular gases such as CO_2, O_2, and N_2 diffuse through the membrane without impediment and always in the direction of the gradient if such exists. This mechanism is known as simple diffusion.

47. (D) The cell's endomembrane system, which initially is continuous with the nuclear membrane, provides for transport of substances throughout the cell. This provides for movement of materials manufactured within the endoplasmic reticulum and Golgi. The system also includes organelles such as plastids, mitochondria, endosomes, and lysosomes.

48. (C) The membrane is conceptualized as a fluid in which numerous protein-based molecules float, identified as the fluid mosaic model. Some of these floating proteins are linked to other peripheral proteins and are known as attached proteins. Some of these embedded proteins are attached to many variations of fatty acids and are called lipid-linked proteins. Two types of these integral proteins associated with transmembrane transport include channel proteins, usually with an α-helix structure, and porin proteins with a larger β-barrel structure.

49. (D) The fluidity and permeability of the cell membrane is affected by the precise structure and content of the various phospholipids and proteins it contains. In general, any variation in content that increases the density of the membrane by allowing closer stacking of its components decreases fluidity. The opposite is also true: the less dense the packing, the greater the fluidity. These variations in structure include the precise structure of the hydrophilic head of the phospholipids, and both the length and degree of saturation of the hydrophobic tail.

50. (A) Very large molecules such as proteins or cell-sized particles cannot pass thorough a phospholipid bilayer without the process of endocytosis. Charged ions and large polar molecules cross a membrane either very slowly or not at all. Small polar molecules can pass through, but usually only with some assistance. Small nonpolar molecules, such as CO_2 and O_2, and water (which is polar) pass through a membrane unimpeded and in the direction of the concentration gradient.

51. (E) There are two basic forms of protein transport within a cell. The first is nonvesicular, which includes any protein manufactured within the cytosol and which lacks any signal sequence directing its association with a membrane. The second is vesicular transport, which includes both membrane-bound and non-membrane-bound associations. In order for a protein to move from one membrane-bound structure, such as the Golgi or endoplasmic reticulum, to another requires vesicular transport.

52. (A) The lipid tails embedded between the phosphate layers of the cell membrane are extremely hydrophobic and physically resist association with any charged particle, which includes ions. In order for these ions to move across a membrane, an insulating tube, commonly seen as a secondary protein structure such as an α-helix, must be present and structurally compatible with the movement of any specific ion.

53. (E) The phospholipid molecules within a membrane are free moving and not bound to any other phospholipid by covalent bonds. This independence allows each of these molecules to rotate, move laterally across the surface of the membrane much like an ice cube might move on the surface of water within a glass, or bend within certain boundaries. However, to move from one leaflet to another, known as flip-flop, requires a large energy expense and is only rarely observed.

54. (B) Charged ions, such as Na^+, Cl^-, K^+, and H^+, cannot cross a phospholipid bilayer without some transport mechanism, regardless of whether it is with or against a concentration gradient. If against a gradient, the transport will always require some energy expense. However, if in the direction of the gradient, then channel-mediated passive transport is adequate.

55. (B) Disulfide bridges will form between the two sulfur-containing cysteines, neither of which are not found in oligosaccharides. These bridges are formed by enzymatic action within the endoplasmic reticulum and Golgi apparatus. When crosslinked membrane-bound proteins are transported to the cell membrane, they are unpackaged when the transport vesicle fuses with the membrane. The fusion mechanism always places these modified proteins in the exterior leaflet of the membrane.

56. (C) Cells are extremely frugal when it comes to using stores of their hard-fought-for energy. Thus when transport across a membrane can be coupled with the force behind a concentration gradient rather than chemical energy, it is an ideal solution. Coupled transport is an energy-efficient solution that exists in two forms: symport, when the transported molecules are moving in the same direction, and antiport, when they are moving in opposite directions.

57. (D) Cell function is commonly linked to cell shape. The shape of eukaryotic cells is maintained by a protein scaffolding that, in animals and protozoans, also assists in cell movement. This scaffolding is always present on the inner (cytosol side) leaflet of the cell membrane and is called the cell cortex.

58. (C) Because of the fluid mosaic structure of membranes, components such as the phospholipids and membrane-associated proteins float freely across one or both of the membrane's leaflets unless stabilized by some sort of internal scaffolding. When the functioning of the cell is very closely associated with precisely defined locations on the membrane, then these can be stabilized with intracellular cortex aggregates.

59. (B) Energy efficiency is key to a cell's survival. Just as a town might use energy to pump water up into a tower for storage so the gravity gradient might be used to later disperse the resource, so a cell will use chemical energy in the form of ATP to pump ions against a gradient to store that energy. Thus later, when force is required to move molecules across a membrane, the energy stored in the gradient can be used efficiently. Sodium ions in animals are the most commonly such used molecules.

60. (D) No substance can cross a membrane against a gradient unless assisted by some cellular mechanism. Additionally, glucose is a rather large uncharged molecule that requires protein assistance regardless of the direction of flow. This means that glucose can move across a membrane against a gradient only with the expense of energy (active transport) with help from large assisting proteins (carrier mediation).

61. (B) Several physical factors dictate the functionality of a membrane. One of these factors is the length of the number of carbon atoms in the tails that comprise the middle hydrophobic layer of the membrane. If the length of these varied widely, then the tails of one layer would interfere with the tails of the other. Additionally, the fluidity of the membrane is affected by the length of these tails. The vast majority of the tails within the functioning membrane are composed of 18–20 sequential carbons.

62. (A) The movement of proteins to the chloroplast (and the mitochondria as well) is by the same mechanism as that used for assembly within the endoplasmic reticulum. The appropriate primary amino acid signal sequences that identify a protein as belonging to the chloroplast are there solely for that purpose and must be excised after delivery and binding of the protein. After translation is completed, the protein refolds and the now unneeded signals are removed.

63. (A) There are three known forms of gating (switching on and off) ion channels in neurons. Some are controlled by physical signals (mechanically gated), some by binding of a neurotransmitter (ligand gated), and some by a buildup of ions (voltage gated). The signal that initiates the formation of an action potential that moves down an axon is usually initiated by the binding of a ligand, with the exception of sensory preceptor cells, but the propagation down the axon is continued by the flood of ions into the neuron.

Chapter 4: Enzymes

64. (D) DNA polymerase and DNA gyrase are both associated with DNA replication, or copying the DNA genome of a cell prior to cellular division. Restriction endonucleases are enzymes that protect bacteria from viral infections. Reverse transcriptase synthesizes a strand of DNA from an RNA template and is thus also known as RNA-dependent DNA polymerase. RNA polymerase synthesizes RNA from a DNA template.

65. (C) Both pH and temperature affect enzyme function by influencing the degree of folding within the three-dimensional structure of the molecule. Cofactors are compounds that serve to activate or accelerate enzyme function; inhibitors, when they are present, do the opposite and serve to reduce enzyme activity. Nitrogen, on the other hand, has no influence on enzyme activity.

66. (B) Enzymes are simply catalytic proteins and are constructed of amino acids. Lipids are composed of fatty acids and can be attached to proteins in the form of lipoproteins but are not essential elements of enzymes. Similarly, simple sugars are used to construct carbohydrates such as starches and cellulose and may be attached to proteins in the forms of glycoproteins or proteoglycans, but they are not essential components of enzymes. Ions may affect enzyme function in the form of cofactors but are not used in the construction of enzymes.

67. (D) An enzyme is a globular protein with catalytic activity. A catalyst is any substance that lowers the energy of activation of a chemical reaction, either anabolic or catabolic, and is not consumed in the reaction. Enzyme activity is sometimes controlled by interactions with cofactors or coenzymes that bind at an allosteric site. The location on an enzyme that actually participates in the reaction is known as the active site.

68. (C) The amino acids in any protein are bound together by peptide bonds, which are a form of covalent bonds; this maintains the primary structure of the protein only. The three-dimensional functional shape of any protein is maintained by hydrogen bonds between nonsequentially adjacent amino acid side groups and by disulfide bridges between cysteines.

69. (A) Enzymes are not the only proteins capable of catalytic function. If an antibody is produced to bind to an enzyme's active site, its own binding site will form a reverse structural image of that active site. If an antibody is then produced to bind to this reverse structure, then the resulting complementary binding site will mimic the structure and the function of the original active site, producing an abzyme. The only other biologic molecule, other than proteins, that has the capability to perform catalytic function is a ribozyme, a form of autocatalytic RNA.

70. (E) Choices A and C define what enzymes do in their most basic form. Enzyme components of a eukaryotic spliceosome catalytically remove introns from heterogeneous nuclear RNA to produce mRNA. The enzyme DNA polymerase catalytically synthesizes DNA from precursor nucleotides. The binding of one cell to another is accomplished by ligand-receptor binding, not by catalysis.

71. (E) Enzymes are involved as catalysts of biologic chemical reactions. As such they might either facilitate the release of energy from breaking apart a compound or storing energy in the formation of a new bond. In all cases the function of an enzyme is to lower the energy of activation of any reaction whether it is exergonic or endergonic.

72. (C) Chemical convention uses word suffixes to identify structure or function. For example, the suffix *-ose* designates a carbohydrate such as glucose or cellulose. The suffix that is used to designate any enzyme is the *-ase* ending of a word. Only choice C lacks this suffix, indicating that it is not an enzyme.

73. (D) Enzymes are very large proteins and their mass is measured in thousands of daltons (kDa). The large molecular mass is required to maintain the proper shape necessary to maintain enzymatic function. Small molecules, with fewer hydrogen bonds between reactive groups, would be much less capable of maintaining this proper shape. Choice D, with a mass of only a few atoms, is much too small to indicate an enzyme.

74. (B) An enzyme's active site must have the ability to reach around and align one or more bonds between adjacent atoms in a substrate or substrates. This interaction allows the enzyme to twist or contort the positioning of the atoms into the most advantageous positions. This means that choices A, C, and D are the least likely. Of the remaining two choices, option E can be eliminated because a very large active site would most likely reduce the rate and function of enzyme activity.

75. (C) Exergonic reactions are those involved in energy-releasing processes. This means that bonds are being broken, and the energy they once contained is being used elsewhere. This process is best linked to catabolic reactions, which comes from the Greek *kata* (downward) and *ballein* (to throw).

76. (D) Enzymes are catalytic proteins composed of scores to hundreds of amino acids, all of which are organic molecules. Thus the best description of an enzyme is a large globular protein with catalytic properties.

77. (B) Enzyme activity can be regulated by inhibitors, which decrease activity, and cofactors, which increase activity. Cofactors may be an inorganic molecule such as Mg^{2+} or Ca^{2+} or an organic molecule such as NAD, in which case it is called a coenzyme.

78. (B) In order to break or form a bond in the presence of a catalyst such as an enzyme, energy must be added to the molecular bond to destabilize it. When sufficient energy has been pumped into the system, the reaction occurs and energy is released. The difference between the initial energy level and the resulting energy level is identified as ΔG, or the difference in free energy. If the energy contained in the new bond is less than the original, then it is identified as a $\Delta G(-)$, while more is a $\Delta G(+)$.

79. (A) Enzyme activity can be reduced by substances known as inhibitors. If the inhibitor binds to the enzyme at a location other than the active site, it is known as a noncompetitive inhibitor. If, however, the inhibitor competes with the substrate for the active site, then it is known as a competitive inhibitor.

80. (E) A cofactor has the opposite effect of an inhibitor on enzyme activity. Whereas enzyme activity is reduced by an inhibitor, a cofactor increases enzyme activity. In fact, if a required cofactor is not present in an enzyme system, then the enzyme is incapable of any catalytic activity, because the active site is in an improper configuration.

81. (D) Exergonic reactions are those that release free energy from molecules. These are primarily observed in catalytic reactions where larger molecules are broken down into smaller components. Endergonic reactions are the opposite, where free energy is stored in the form of chemical bonds, usually in the processes of building larger molecules from smaller subunits in anabolic processes. Thus lipid synthesis is endergonic, not exergonic.

82. (C) Chemical reactions are processes where initial compounds interact to form final compounds. In enzymatic processes, these initial forms are called substrates and the final forms are called products. In inorganic forms, the equivalents are called reactants and products. Thus, the terms *reactants* and *substrates* are most analogous.

83. (E) Enzymes function by lowering the energy of activation required to reform bonds. A substrate enters into the active site of an enzyme, and the shape of the enzyme changes to induce stress in the substrate bond being affected. However, without the addition of enough energy to reach the required activation level, the catalytic properties of the enzyme cannot function.

84. (D) Reduction is any process where a molecule either accepts an electron or a hydrogen ion or loses an oxygen atom. When one molecule is reduced, another is always also oxidized, as the reactions are coupled. Thus oxidation is when a molecule either loses an electron or a hydrogen ion or gains an oxygen atom.

Chapter 5: Catabolic Metabolism

85. (E) One two-carbon atom molecule (acetyl CoA) from glycolysis will enter the TCA cycle by attaching to a recycling molecule with four carbon atoms (oxaloacetate) to produce a molecule with six carbon atoms (citrate). The cycle continues with a molecule of CO_2 being released, producing a molecule containing five carbon atoms (α-ketoglutarate), which in turn releases another molecule of CO_2 to produce a four-carbon-atom-containing compound (succinyl CoA).

86. (A) Anabolic reactions synthesize larger molecules from smaller ones and store energy within the molecular bonds. Catabolic reactions are the opposite and reduce larger molecules to smaller ones, thereby releasing the energy stored within their bonds. Metabolism is the summation of all catabolic and anabolic processes in a cell.

87. (A) Oxidizing organic compounds release energy. The unit of measure of this energy is the calorie (c), which is the amount of heat it takes to raise 1 gram of water 1 degree centigrade. Because this is rather small, we more commonly use the unit kilocalorie (kcal or C). Both carbohydrates and proteins contain approximately 4 C per gram. Lipids, on the other hand, contain more than twice that at 9 C per gram.

88. (C) Adenosine triphosphate (ATP) contains three phosphate groups in sequence. Each of those phosphate groups contains a high-energy electron that can be transferred to other molecules for use in bond destabilization. Thus ATP contains more energy than ADP, which contains more than AMP or cAMP. NADH similarly contains a single high-energy electron. However, when a molecule of glucose is fully oxidized through aerobic respiration, it results in the production of 36 ATP (eukaryotes) or 38 ATP (prokaryotes).

89. (D) In order to harvest energy from a molecule of glucose, the molecule is subjected to a sequential series of enzymatic steps that splits the six-carbon-atom-containing glucose into two identical three-carbon molecules. It takes five steps to convert glucose into two molecules of phosphoglyceraldehyde (PGAL) and then an additional five steps to convert the two PGAL molecules into the end product of glycolysis, pyruvate.

90. (B) Organisms can be classified based on their nutritional requirements for energy, carbon, or electron sources. An organism can get its energy from either the sun (phototroph) or chemicals (chemotroph). An organism can get its carbon in inorganic form from the air or water (autotroph) or in prepackaged organic form (heterotroph). And an organism can get its electrons from either inorganic (lithotroph) or organic (organotroph) sources. Thus an organism that gets both its energy and carbon from the same molecule would be identified as a chemoheterotroph.

91. (E) The mitochondrial electron transport chain consists of both protein and nonprotein electron carriers physically arranged in sequence within the mitochondrial membrane. These include the flavoprotein FMN, derived from vitamin B_2; ubiquinone, a nonprotein component that is also known as coenzyme Q; and the iron-containing cytochromes. Pyruvate is the end product of glycolysis and is not an ETC component.

92. (D) After NADH is oxidized back into NAD, the first component of the electron transport chain, FMN, is reduced with the donated high-energy electron. After passing down through the ETC components, the energy of this electron is drained off to power the proton pumps that move hydrogen ions across the mitochondrial membrane. In order for another electron to pass down through the system, the previous drained electron must be discarded onto a final electron acceptor sink. This sink component is oxygen gas, which is reduced to a molecule of water.

93. (B) Chemotrophic organisms use the energy contained in chemical compounds for driving their metabolic processes, as opposed to phototrophic organisms that use the energy of light for the same reason. The term *aerobic* refers to the presence of oxygen molecules that act as terminal electron receptors required by oxidative phosphorylation and is independent of energy sources.

94. (E) All four of the options are components of the citric acid cycle, also known as the tricarboxylic acid (TCA) cycle or the Krebs cycle. In this process, citrate, containing six carbon atoms, is converted to the five-carbon-containing α-ketoglutarate, releasing a molecule of CO_2 in the process. Similarly, the α-ketoglutarate is converted to the four-carbon-containing succinyl CoA. After that there are several molecular rearrangements until oxaloacetate is produced, which then combines with a molecule of acetyl CoA to restore a molecule of citrate for the continuation of the process.

95. (C) A molecule of glucose contains six carbon atoms and the covalent bonds between each of them contains stored energy. A molecule of pyruvate contains three carbon atoms. The production of two molecules of pyruvate from a single molecule of glucose is called glycolysis, which is a catabolic process identified by the reduction in size of organic molecules and the resulting release of free energy.

96. (C) The energy contained within a molecule of glucose is most efficiently produced when harvested through several mechanisms. These mechanisms include glycolysis, pyruvate conversion to acetyl CoA, and the electron transport system (ETS) housed within the mitochondria. Although some initial ATP investment is required to initiate the process, the net production is 36 ATP in eukaryotes when in the presence of oxygen. The lack of oxygen prevents the functioning of the ETS, which reduces the ATP yield to 2. Prokaryotes are slightly more efficient and can produce 38 ATP from a single molecule of glucose.

97. (B) Terms used to describe the nutritional profiles of various organisms include prefixes that identify their source of electrons (*litho-* vs. *organo-*), their source of carbon (*auto-* vs. *hetero-*), and their source of energy (*photo-* vs. *chemo-*). Thus a photoheterotroph identifies an organism that feeds (*-troph*) on the energy of light (*photo-*) and on carbon from other (*hetero-*) organisms.

98. (E) Fatty acids are composed of long chains of carbon atoms and are used in the construction of mono-, di-, and triglycerides. Most of the fatty acid chains found in cells contain an even number of these carbons. Fatty acids are usually tapped as an energy source only when other forms of chemical energy storage such as ATP, carbohydrates, and proteins have been used. Pairs of carbon atoms are removed, converted to acetyl CoA, and then fed into the Krebs cycle in a process known as β-oxidation.

99. (B) Whenever oxygen is added to a molecule, or when an electron (e^-) or hydrogen ion (H^+) is removed, the process is called oxidation. The opposite processes of removing an oxygen atom or adding an e^- or H^+ is called reduction. Oxidation and reduction processes are always coupled and are known as redox reactions. The conversion of NADH to NAD involves the removal of a hydrogen ion and is thus an oxidation reaction.

100. (A) Organisms that are chemoautotrophs use chemical compounds as an energy source as opposed to using energy extracted from photons. These organisms, as autotrophs, use inorganic carbon molecules as their carbon source as opposed to the preformed organic compounds required by heterotrophs. Carbon dioxide is a molecule generated as a waste product in respiration because it no longer contains enough energy to be useful to organisms and is thus not used as an energy source.

101. (D) When an electron is delivered by NADH or $FADH_2$ to the early components of the electron transport system, whether found in eukaryotic mitochondrial or prokaryotic plasma membranes, it contains enough energy to power two or three proton pumps in sequence. However, once this energy has been significantly depleted from the electron, it is then passed through the later cytochrome components. Their job is simply to bleed off the remaining energy until it can be discarded onto a molecule of oxygen, converting it to a molecule of water. One of the proton pumps is coenzyme Q.

102. (E) Cellular respiration operates under two different conditions: aerobic (in the presence of oxygen) or anaerobic (when oxygen is not present). Aerobic respiration is much more efficient, allowing the cell to convert a molecule of glucose into CO_2, water, and 36 ATP molecules. Anaerobic respiration, which does not permit the use of the electron transport system, has a much lower energy yield (2 ATP per glucose molecule) because much of the energy remains in organic molecules. This process is also known as fermentation.

103. (E) The energy of glucose is released in three major steps: glycolysis, pyruvate conversion, and the electron transport system. The latter two occur within the mitochondria of eukaryotic cells. Glycolysis, the 10-enzymatic-step process of breaking down glucose into pyruvate, takes place within the cytosol of both prokaryotic and eukaryotic cells.

104. (D) The process of oxidizing a single molecule of pyruvate, containing three carbon atoms, into a single molecule of acetyl CoA, containing only two carbon atoms, generates a single molecule of the waste gas CO_2 and the reduction of a molecule of NAD to NADH. The high-energy electron contained within the NADH is then transferred to the electron transport system, where the energy of the electron is used to pump six pairs of hydrogen ions across the mitochondrial membrane. Each of these pairs, in turn, flood through a molecule of ATP synthase, and this energy drives the conversion of ADP to ATP. Three pairs of H^+ thus produce three ATP molecules.

105. (A) The TCA cycle starts when a molecule of acetyl CoA, containing two carbon atoms, combines with a molecule of oxaloacetate, containing four carbon atoms, to produce a single six-carbon-containing molecule of citrate. The remainder of the cycle drains the energy contained within the bonds of those two new carbon atoms, which are then discarded as waste CO_2 molecules. Thus each cycle brings in two carbon atoms and discards two carbon atoms.

Chapter 6: Photosynthesis

106. (E) The Calvin-Benson cycle is also known as the light-independent pathway of photosynthesis. In this cycle, carbon dioxide from the atmosphere is fixed into organic form and used to synthesize glucose. Of the choices provided, this cycle is not involved with photolysis, which is a portion of the light-dependent pathway.

107. (A) When a photon of light interacts with a pigment molecule, it passes its energy to a valence electron. This electron then moves to an energy level farther from the nucleus than it was in the ground state. Because this energy level is unstable, the electron might leave the pigment, causing its oxidation, and pass on to another molecule, causing its reduction. The alternative is to return to its ground state by releasing the energy in the form of a photon, but this photon will always be of a longer, lower-energy wavelength.

108. (D) A pigment is any substance that can absorb the energy contained within a photon of light. We commonly see color within a pigment based on the energy levels of the photons that are reflected rather than absorbed. Chlorophyll appears green because it is capable of absorbing the energy from photons that we perceive as red, orange, yellow, blue, indigo, and violet. Carotinoids are similar pigments, but they can absorb only the shorter wavelengths at energy levels above green, so we see them as red, orange, and yellow.

109. (C) Photons are packages of electromagnetic energy that have characteristics of both particles and waves, yet have no mass. As they exist as part of the electromagnetic spectrum, each photon can be measured as some specific wavelength, and the shorter the wavelength, the greater the energy content. Thus the seven colors best associated with the rainbow, when arranged in increasing energy content, are red, orange, yellow, green, blue, indigo, and violet.

110. (A) Using the energy in a photon of light to generate a molecule of ATP is called photophosphorylation, which may use either a cyclic or noncyclic mechanism. Noncyclic photophosphorylation involves the removal of electrons from water by photolysis and their transport through photosystems I and II in the chloroplast to eventually produce ATP. In this case, the electrons eventually are picked up by NADP to produce NADPH. In the alternate cyclic form, the electrons are returned to their molecule of origin.

111. (B) In eukaryotes, photosynthesis occurs within chloroplasts, which are similar in size to mitochondria. The chloroplasts contain stacks of disk-like objects connected within a membrane-bound matrix called grana. Each disk is a membranous structure containing chlorophyll that acts as an antenna to harvest the energy of passing photons and is thus the smallest of the structures listed.

112. (D) The process known as photosynthesis consists of two parts: the light-dependent reactions (formerly known as the light phase) and the light-independent reactions (formerly known as the dark phase). In the former, the energy of light is harvested and converted into the form of ATP. In the light-independent reactions, the ATP is used to drive a series of anabolic steps that result in the production of glucose using inorganic carbon fixed from the atmosphere. This synthesis process is known as the Calvin-Benson cycle.

113. (C) Noncyclic photophosphorylation involves the light-capturing processes found in the chloroplasts of eukaryotes. The energy of captured photons splits molecules of water, releasing oxygen gas, hydrogen ions, and the electrons that are transported through a series of electron-carrying proteins known as photosystems I and II (also known as P680 and P700). These eventually are used to reduce NADP to NADPH. The analogous simpler process found in photosynthetic bacteria is known as noncyclic photophosphorylation.

114. (A) Ribulose bisphosphate (RuBP) is a five-carbon-atom-containing molecule contained within the stroma of the chloroplast. The energy present in the form of ATP produced in the light-dependent reactions is used to drive the reactions that attach an inorganic carbon from a molecule of carbon dioxide onto the RuBP. The molecule is split into two molecules of phosphoglycerate of three carbon atoms each, which are then used to eventually reconstruct RuBP. After six carbon fixation cycles, a molecule of glucose is released. This glucose-producing pathway is known as the light-independent reactions and the Calvin-Benson cycle.

115. (E) Oxygen is vital to eukaryotic cells because it serves as the final electron acceptor for the mitochondrial electron chain. Without oxygen, the electrons can no longer flow through the chain, the ATP required for cellular operations can no longer be synthesized, and the cell dies. This oxygen is made available by photolysis.

116. (E) Photosynthesis is a process in which prokaryotic or chloroplast-containing eukaryotic cells use the energy harvested from photons to drive anabolic processes. These processes are conducted by autotrophic, not heterotrophic, cells that have access to light. The harvested energy is used to convert water and carbon dioxide into sugars such as glucose.

117. (B) Any molecule that interacts with a photon of light must have valence electrons that can absorb the proper quantum of energy represented by that photon. That energy level is perceived as the color (wavelength) of that light. Pigment molecules such as chlorophyll contain such electrons and are normally detected visually by the colors that are not absorbed and are reflected away from the pigment.

118. (D) *Phototroph* literally means "feeding on light" and refers to organisms that use the energy originally contained in photons to drive metabolic processes. These organisms most commonly function independently of preformed organic compounds and thus are classified as autotrophs, literally meaning "feeding oneself." This usually, but with some prokaryotic exceptions, means that most phototrophs are also autotrophs.

Answers 135

119. (A) Phototrophs possesses the molecular machinery that allows them to be photosynthetic. Nonphototrophs are heterotrophs, literally "different feeders," meaning that their energy comes from other organisms in the form of organic molecules such as glucose, lipids, and proteins.

120. (B) Most phototrophs are also autotrophic, but there are prokaryotic exceptions. A photoheterotroph would be an organism that uses light as a source of energy but requires preformed organic molecules as its carbon source, instead of fixing carbon from the atmosphere to satisfy all of its organic compound needs. These organisms are commonly oxygenic, releasing oxygen as a metabolic waste product, rather than aerobic, requiring oxygen for respiration.

121. (D) The thylakoid membrane is a structural component of the grana contained within a chloroplast. Within this membrane are all of the molecular components that drive the light-dependent reactions of photosynthesis. The cytochromes are iron-containing proteins that permit the transfer of electrons through the electron transport system of the mitochondria.

122. (C) Bacteriorhodopsin is a photosynthetic pigment produced in a number of Archaea, most notably of the salt-loving genus *Halobacterium*. This pigment allows the organism to harvest the energy of light in a process known as cyclic photophosphorylation, as opposed to the noncyclic form found in chloroplasts. This pigment is purple, although this genus is not classified within the purple sulfur or purple nonsulfur bacteria.

123. (D) In order for the light-dependent reactions of the noncyclic photophosphorylation pathways to occur, a source of electrons must be tapped. These electrons are then pumped to higher energy levels by the absorption of the energy contained in photons of light. This process is known as photolysis, which means the "breaking apart of water."

124. (A) Molecules can absorb the energy of light if they have valence electrons at the proper quantum level, meaning that many compounds interact poorly with light or not at all. Compounds that can absorb light energy will do so at selected wavelengths, which can be perceived by the wavelengths they do not absorb, with white reflected light indicating no absorption at all. Pigments are compounds that we perceive as having colors based on their absorption patterns.

125. (D) The type of photophosphorylation that occurs in the chloroplasts of eukaryotes is noncyclic in form because the final acceptors for electrons transferred for energy production are not the same as the compounds they came from. In the more primitive form of cyclic photophosphorylation, the energized electrons are removed from their origin but then immediately returned as the energy is harvested. This form of energy capture is found only in prokaryotes, thus grana are not present.

126. (D) Photosystems I and II present in the mitochondrial membranes function by sequentially transferring electrons from component to component. These electrons come from the breakdown of a molecule of water, releasing oxygen gas as a waste product, hydrogen ions to generate ATP through substrate-level phosphorylation, and the needed electrons.

Chapter 7: Mitosis

127. (E) Mitosis is the form of asexual reproduction in eukaryotes and is a highly regimented and controlled process. After all of the various steps have been completed, the last phase is the actual reconstruction and separation of the two new nuclei, a process known as cytokinesis.

128. (E) Interphase is the portion of the cell cycle in which most cells normally reside. During this phase, the cell is maintaining homeostasis, metabolizing to provide energy and materials for that homeostasis and growth, and waiting for the next mitotic process. However, the division of some cells could be detrimental to some organisms where cellular constancy is more important than growth. This includes neurons that store memories and thinking processes within a physical structural network of connections and lymphocytes that provide unchanged immunologic memory spanning decades.

129. (C) Diploid cells have two versions of every chromosome, which are mostly, but not completely, identical in gene positions and size, as one version comes from the father and one from the mother. The actual number of chromosomes varies from species to species, and this number is commonly designated with the letter n. Thus the chromosomal content of a normal diploid cell is designated with the term $2n$. The cell cycle starts with a diploid cell ($2n$) in interphase, passes through the various portions of M phase, resulting in the duplication of the DNA in the nucleus (now $4n$), and then undergoing cellular division, resulting in the formation of two diploid cells ($2x2n$).

130. (B) Metaphase is the portion of M phase in which all of the duplicated chromosomes align along the equatorial plane equidistant from the centrioles present at opposite poles of the cell. M phase is the portion of the cell cycle that includes all the phases of mitosis.

131. (C) There are two forms of cellular reproduction in eukaryotes. One of these, meiosis, is found only within the gonads of multicellular organisms and results in the formation of four genetically different haploid cells called gametes. The other most common form, mitosis, results in the formation of two genetically identical daughter cells.

132. (C) Both trisomy and transformation indicate conditions or processes that are genetically abnormal. Reverse transcriptase, also known as RNA-dependent DNA polymerase, is the enzyme that is the key for the replication strategy of human immunodeficiency virus (HIV). Meiosis is the form of cellular division associated with sexual reproduction, while mitosis is associated with the asexual process.

133. (B) The third stage of mitosis is anaphase. It follows metaphase when the chromosomes align on the cellular equatorial plane. Next, in anaphase, the protein filaments attached to both the chromosomes and the centrioles at the cell poles shorten, in effect reeling the separating chromosomes toward opposite ends away from the central plane.

134. (B) During interphase in the cell cycle, the DNA within the nucleus is in a more dispersed form called chromatin. This relaxed condition permits transcription factors to interact with the DNA, allowing the cell to express genes and produce products necessary for maintaining homeostasis. At some point most cells will initiate M phase with the transcription of the proteins required for cellular division. When these factors are completed, mitosis is initiated with the condensation of the chromatin into the more densely packed form called chromosomes during prophase.

135. (D) DNA replication requires the action of multiple enzymes, including topoisomerases and helicase. Transposase is an enzyme associated with the relocation of mobile genetic elements. Recombinase is often associated with DNA repair and related functions. Telomerase is the enzyme responsible for repairing or extending the telomeres at the ends of each chromosome, which act to prevent the degradation of coding portions of the DNA. Recently, telomerase activity has also been linked to cancer expression.

136. (E) Ribosomes are responsible for the cellular process of translation when proteins are synthesized based on the triplet code found on transcribed mRNA. If ribosomes could not function, then translation would cease and no new proteins could be synthesized. Since the synthesis of mitotic proteins is necessary in order for mitosis to occur, cell replication would cease immediately.

137. (E) Normal asexual cellular reproduction is when an initial mother cell duplicates its own DNA with great fidelity and then divides in such a way as to result in two genetically identical daughter cells. This means the DNA content, distribution, and genetic sequences of the two daughter cells are not only identical to each other, but also identical to the original mother cell of which they are clones.

138. (A) The word *cytokinesis* comes from the Greek roots *cyto-* (meaning "cell") and *kine-* (meaning "movement"). In biology, it applies generally to the movement of cells away from each other following cellular division. More specifically, it applies to the closure of the plasma membrane, which results in the separation of the two new daughter cells.

139. (A) Cellular division is different when comparing the processes of prokaryotes and eukaryotes. In the eukaryotic cell cycle, some processes shut down, while others initiate to allow synchronization of processes. Bacteria, on the other hand, go full bore at all times in a much less controlled manner. Therefore, they lack the cell cycle of eukaryotes.

140. (C) Mitosis is a process that produces two identical daughter cells from an original mother cell. This identity is not only genetic but also includes the distribution of all organelles and cytosolic contents.

141. (B) Eukaryotes have vastly larger genomes and more complex chromosomal organizations when compared to prokaryotes. They thus require a replication process that proceeds with a rigidly controlled step-by-step mechanism to ensure accuracy and fidelity. This regimented process is known as mitosis, while the much simpler replicate-then-divide process of prokaryotes is called binary fission.

142. (E) S phase is the synthesis portion of the eukaryotic cell cycle during which the complete genome is replicated. At the beginning of S phase, the cell contains $2n$ DNA, and at the end of S phase, it has $4n$. Because the amount of genetic material changes during the S phase, the exact distribution to the daughter cells also changes and is not determined until the cell cycle continues into the G_2 phase.

143. (B) Chromosomes contain the genetic makeup that determines the phenotype of any specific eukaryotic organism. However, because of the massive size of the DNA molecule enclosed, the entire lengthy molecule is carefully organized into much smaller subunits called nucleosomes. Each nucleosome consists of about 146 base pairs of DNA wrapped around a spool of histone proteins of approximately 12 nm in diameter.

144. (E) Sister chromatids are formed by the exact replication of each individual chromosome prior to both mitosis and meiosis. They are kept in close proximity by proteins that join them at a region called the centromere. Spindle fibers attach to the centromere via kinetochores prior to separation during mitosis.

145. (C) Both an ovum and a sperm are formed by the process of meiosis, and both are haploid cells classified as gametes. When the ovum is fertilized by the sperm, a new genetically unique cell is produced. This fertilized ovum is also known as a zygote, which after undergoing several mitotic divisions becomes an embryo.

146. (B) For mitosis to form two genetically identical daughter cells, the replicated DNA of the mother cell must be separated in an identical manner. The centrioles involved in separating the sister chromatids equally position themselves at opposite poles within the cell undergoing mitosis. After all of the sister chromatid pairs have become aligned on the equatorial plane during metaphase, they separate, drawn by the spindle fibers, during anaphase.

147. (A) DNA replicates during the S phase of interphase. Following the S phase, the cell prepares for the actual separation of the mother cell's DNA to two daughter cells by manufacturing the separating machinery during G_2 phase. Once available, the cell initiates the M phase by dissolving the nuclear membrane and condensing chromatin into the more visible chromosomes during prophase.

Chapter 8: Meiosis

148. (A) There are two forms of eukaryotic cellular division. The asexual form involving mitosis converts an initial mother cell into two identical daughter cells. The sexual form in which an initial mother cell converts into four genetically distinct haploid cells is called meiosis.

149. (A) While the goal of mitosis is genetic fidelity, the goal of meiosis is genetic variation. The purpose of mitosis is to produce exact replicas, or clones, of the original mother cell, each capable of all life functions to include replication. The resulting cells formed by meiosis, however, have lost their ability to undergo independent replication, because they contain only half of the amount of genetic material needed for that process and, in the case of sperm, have an extremely limited lifetime.

150. (B) While mitosis involves a single division following DNA replication, meiosis involves two. The mechanisms involved in the first division are identified as meiosis I and in the second division as meiosis II. Other than the crossing-over event that occurs during meiosis I, the division process is nearly identical to mitosis, and both result in a DNA distribution of $2x2n$.

151. (D) Meiosis begins with a diploid mother cell containing $2n$ DNA. Prior to meiosis I, the DNA is replicated during the S phase and becomes $4n$. The first reduction division divides the DNA into two cells, each with $2n$ DNA, and the second, meiosis II, completes gametogenesis with the formation of four haploid cells ($4\times 1n$).

152. (D) During prophase I, which is the prophase portion of meiosis I, the mother cell's nuclear membrane dissolves, chromosomes condense, and tetrads align and undergo genetic mixing in the form of crossing over. During metaphase I, the chromosomes all align on the equatorial plane. It is during the following phase, anaphase I, that the chromatids segregate to the opposite poles of the cell.

153. (B) While mitosis focuses on keeping the DNA unchanged from generation to generation, meiosis focuses on increasing genetic variation. One of the major components of obtaining this variation is with the crossing over of matching segments of chromosomal DNA between homologous chromosomes. In order to make sure there is an even exchange of the same portions of the chromosomes, they align gene-to-gene in synapsis during prophase I.

154. (A) Sexual reproduction involves the fusion of two gametes from the same species. These gametes are designed to produce maximal genetic variation while keeping the phenotypic constancies of the species. Of the two forms of reproduction found in eukaryotes, the form involved in sexual reproduction is called meiosis.

155. (D) Meiosis involves two reduction events following the replication of DNA during the S phase. The first division produces two genetically distinct diploid daughter cells and the second, which occurs immediately following the first, reduces each of those two distinct cells into four even more distinct haploid cells, thus resulting in a DNA distribution of $4\times 1n$.

156. (E) Through the processes of gene duplication and random mutations, new gene functions can be formed, as well as the production of multiple alleles, or versions of genes. Crossing over during meiosis increases the genetic mixing of these alleles and ensures increased genetic variations in offspring as a result of fusion of varied gametes during fertilization.

157. (D) During mitosis the nuclear membrane, which disappeared during prophase, is reformed at the end of telophase. Similarly, during meiosis, the nuclear membrane reforms at the end of the total process at the end of telophase II. However, the membrane also reforms at the end of telophase I of meiosis I, although it immediately disappears again as meiosis II commences.

158. (D) Meiosis is a form of reproduction that increases genetic variation within eukaryotic species. Meiosis results in the production of haploid gametes, which later fuse to produce a diploid zygote. Meiosis occurs as a primary form of reproduction in plants and animals, although many plants and some animals are also capable of parthenogenesis or apomixis. Additionally, it is a common process found in many fungi and some protozoans.

159. (B) Crossing over is a genetic mixing mechanism that occurs during the prophase I portion of meiosis. *Cytokinesis* is the term used to describe the reformation of the nuclear membrane and closure of the plasma membrane following both mitosis and telophase II of meiosis.

160. (C) The major difference between mitosis and meiosis is that the goal of the former is genetic constancy, while the goal of the latter is genetic variation. Besides the crossing-over mechanism, random distribution of chromosomes during meiosis, and sexual selection and reproductive mechanisms, genetic variation is also increased by recombination of alleles by recombinase enzymes.

161. (A) Gametes are genetically reduced cells produced by meiosis for the purpose of sexual reproduction. This process occurs in all four of the eukaryotic kingdoms, which includes plants. The fusion of gametes, which produces a zygote, is called fertilization.

162. (A) During meiosis I, the mechanism observed during mitosis is replicated with the exception of the genetic exchange that occurs during prophase I. Meiosis I ends with the formation of two diploid cells that are genetically unique. Prophase II begins immediately after the completion of meiosis I with the same processes observed with prophase of mitosis, which includes the condensation of chromatin to chromosomes and the dissolution of the nuclear membranes of both new cells.

163. (C) A cell that is heterozygous for gene D will contain one copy of allele D and one of allele d. Meiosis I will result in two cells, both of which may be heterozygous or homozygous for either allele because of crossing over. Meiosis II will result in the formation of two gametes containing the D allele and two containing d. The chance of receiving one or the other is equal.

164. (E) Sperm cells are gametes that are haploid, formed as a result of random arrangements of genes during meiosis. They differ from ova, or eggs, in that they contain little or no cytosol and generally are motile, whereas the ova contain cytosol and are nonmotile. The term *gametogenesis* means the formation of gametes, which includes the formation of sperm (spermatogenesis) and the formation of ova (oogenesis).

165. (E) Prokaryotes are considered haploid cells. This is because after cellular division by binary fission, each daughter cell is a clone of the originating mother cell that contains a single circular double-stranded genome. However, under ideal conditions and because of the lack of a cell cycle, each cell is continually undergoing DNA replication, so the time in which the cell is truly haploid is minimal. The only form of DNA mixing in prokaryotes is conjugation, which occurs in only some species, and transformation, which commonly involves the transfer of nongenomic DNA.

166. (B) The cell that is formed as a result of fertilization is diploid because it contains the DNA from both haploid gametes. This cell is known as a zygote. The zygote begins a number of rapid cellular divisions, often with little or no cellular activity other than mitosis occurring. When the zygote becomes multicellular, it becomes known as an embryo.

167. (E) Cells that have just completed a cellular division enter phase G_0, or the resting phase. They remain in G_0 until conditions signal the need for entering either meiosis or mitosis, depending on whether the cells are of gonadal or somatic origin, respectively. Before cellular division can occur, the numerous enzymes required for DNA replication must be manufactured. After the replication machinery is assembled during G_0, the cell then enters the S phase when DNA synthesis occurs.

Answers ‹ 141

168. (C) During the S phase, a cell completes replication of its genomic DNA and has become tetraploid. The next requirement before initiating cell division is the synthesis of the meiotic apparatus. This final step prior to meiosis itself occurs during the G_2 phase.

Chapter 9: Inheritance

169. (C) The ability to form blood clots when properly triggered is an essential part of the ability to maintain homeostasis in animals. This process has both cellular and molecular components, and deficiencies in either can result in uncontrolled bleeding, or hemophilia. One of the most common forms of hemophilia is the genetic lack of a functional protein called factor VIII.

170. (E) When a person is identified as having a certain blood type, their phenotype is being referred to. Persons with type A blood have the A antigen present on their erythrocytes, and the same is true for type B persons having the B antigen. Their genotype, however, can be either AA or AO (or BB or BO). Using a Punnett square for all four possible crosses (AA × BB, AA × BO, AO × BB, AO × BO), it can be determined that all four phenotypes can result from a blood type A × B cross.

171. (B) Ion gradients across a cell membrane are commonly used to power many cellular functions associated with homeostasis, thus the need for tightly controlled ion flow through channel proteins. One genetic abnormality that results in improper control of sodium ion flow can result in an uncontrollable production of mucus in the lungs. This is seen in the affliction we call cystic fibrosis.

172. (C) One of the benefits of random mating is the increase of genetic diversity within a population, which minimizes lethal recessive traits. Conversely, inbreeding increases the likelihood of the continuance of such traits. This was seen historically with the increased presence of the sex-linked recessive trait of hemophilia in the royal houses of Europe.

173. (C) The term *genotype* refers to the presence of specific alleles in an organism's genome. By convention we use capital letters to indicate dominant alleles and small letters to indicate those that are recessive. Because somatic cells have two copies of each chromosome, and each has an allele for every gene, we see two letters representing the alleles. The only exception for humans is the sex chromosome where a male may be hemizygous for X-linked genes, such as *R*–.

174. (C) The two most dominant antigens present on human erythrocytes are the ABO and Rh antigens. The Rh antigens get their name from their initial discovery on the erythrocytes of Rhesus monkeys. While the ABO antigens are expressed codominantly, simple dominance controls the Rh gene expression. The possible phenotypes Rh^+ and Rh^- are controlled by the alleles *R* and *r* respectively.

175. (D) A person with type A blood may be either homozygous (AA) or heterozygous (AO), both of which produce the A antigen on their erythrocytes. Thus all four cross possibilities must be considered. Since B is not present, no combination can result in a child's blood containing B antigen, and a heterozygous cross can result in either a child with type A (75%) or O (25%) blood.

176. (E) Humans rely on diet as their source of amino acids and have metabolic pathways that permit the breakdown of any excess amino acids. Failure in any of these recycling pathways results in the buildup of their associated amino acid, and these buildups can result in distortions in other processes. One such condition has a buildup of phenylalanine that can result in the interference of brain development as seen in phenylketonuria.

177. (B) The cross of a homozygous dominant individual (*RR*) with a homozygous recessive individual (*rr*) will result in every progeny having an equal distribution of both alleles, meaning that every descendent will be heterozygous (*Rr*) and will express the dominant phenotype.

178. (C) Hemoglobin is the oxygen-carrying protein found in erythrocytes. Mutations at certain restricted sites can result in a failure of this protein, which not only interferes with the carrying function but also in the structure of the cell itself. One such mutation produces the disease we call sickle cell disease.

179. (E) Diploid organisms, such as humans, carry at least two copies of every allele distributed evenly on both sister chromosomes (one inherited from the father, one from the mother). The only exception to this is when considering the sex chromosomes X and Y. While females will still have at least two copies of every gene found on their X chromosomes, males will have only one. Any gene found on the X chromosome in males is identified as hemizygous as they have no second X chromosome. We indicate this as R^-, or, in the case of any X-linked recessive trait, as r^-.

180. (D) Normal hemoglobin maintains its tertiary structure regardless of whether it is carrying oxygen or not. Some mutations, however, cause a change in shape and function based on the presence or absence of oxygen. Under conditions of low oxygen content, one such mutation results in the folding, or sickling, of erythrocytes, which impairs their ability to move through capillaries, resulting in pain, oxygen starvation for the respective tissues, and even death. This attack is known as a sickle cell crisis.

181. (B) The ABO blood group system identifies four phenotypes: blood types A, B, AB, and O. There are three codominant alleles that control the ABO expression. When a person has the A allele, he or she will produce the A antigen that is carried on the erythrocytes, and in the same way, the B allele will produce the B antigen. If a person has AB blood, it is because he or she has both the A and B alleles. The third allele, O, produces neither the A nor the B antigen.

182. (B) The term *X-linked* refers to genes present on the X chromosome, meaning they are not found on the autosomes but on one of the sex chromosomes. If a gene is found on the X chromosome and is dominant, it will always be expressed in both males and females. However, an X-linked recessive gene will always be expressed in males, as there is no other allele to dominate its expression, but in females it will be expressed only if homozygous.

183. (A) In order to maintain homeostasis, every metabolic function must be tightly regulated. Failure to regulate any function, such as might occur when a mutation impairs the normal function or structure of a protein, can result in the failure of the entire associated regulated pathway. When one lipid production pathway loses the function of one such regulatory protein, it results in the overproduction of lipids, which then interferes with nerve transduction and the loss of muscle stimulation and mass.

184. (E) When chromosomes were first stained, it was observed that every cell within the same organism contained the same chromosomal patterns. Analysis of these then identified the concept of diploidy and the presence of two of each homologous chromosome. These chromosomes were then identified based on visible differences in length, banding patterns, and centromere position. Later studies also identified that genes could be mapped to specific chromosomes.

185. (A) Chromosomes that are present in all cells as homologous pairs are referred to as autosomes. Those not necessarily in pairs, such as seen in human males, are identified as sex chromosomes. Thus the allele combination of *Rr* indicates heterozygosity (one of both alleles), which may occur either on autosomes or in females on the X chromosomes.

186. (D) The number of genes expressed in an organism is not related to the number of chromosomes present in the nucleus. Human somatic cells contain 23 pairs (46 total) of chromosomes, 22 of which are autosomes with the final pair being the sex chromosomes. Gametes, found only in the gonads, are haploid and contain only one of each chromosome for a total 23.

187. (E) Aspartame is basically a dipeptide, where the two amino acids present are phenylalanine and aspartic acid. In the vast majority of the population, when this molecule is ingested, the sweetener is broken down by peptidases and the amino acids are recycled and incorporated into new proteins. Some, however, lack the ability to break down the phenylalanine, and its buildup interferes with brain development during infancy and through adolescence. This condition of phenylketonuria is controlled by a dietary restriction on protein intake.

188. (A) Sickle cell anemia is an autosomal recessive condition that presents as erythrocytes either full (if the person is homozygous) or half full (if heterozygous) of mutated and dysfunctional hemoglobin. A sickle cell crisis is precipitated when oxygen levels in the tissues drop as a result of slight (in the former) or moderate (in the latter) exercise. The heterozygous condition, which causes sickle cell trait, can be disabling. The homozygous condition, which causes sickle cell disease, can be fatal.

189. (D) The letters found on the top and side of Punnett squares represent the allele, or combination of alleles, that a gamete might contribute to a newly formed zygote. The number of resulting possibilities is based on the number of alleles, with the most common being a single gene represented by two alleles. Thus with two allele possibilities on top and two on the side, the number of possible combinations is four. If two genes are considered, each gamete has four possible combinations, and $4 \times 4 = 16$ possible results. In the same way, with three genes, the number of possible gamete combinations is 8, and $8 \times 8 = 64$.

Chapter 10: DNA Replication and Repair

190. (C) Before the mechanism of DNA replication was worked out by researchers, there were two possible models envisioned. The first was the conservative model, in which the original template strands of the DNA double helix were retained in one daughter cell, while the newly synthesized double helix was transferred whole to the other daughter cell. The second was the semiconservative model in which each daughter cell received one strand from the original template and one newly synthesized complementary copy. We now know that the latter is the actual mechanism.

191. (B) Ultraviolet (UV) light is considered ionizing at the shorter wavelengths because each photon contains sufficient energy to knock valence electrons away from their original atom, thus creating a chemically reactive ion. These reactive ions tend to form chemical bonds with adjacent atoms. When this occurs within a strand of DNA, the most common reaction is the formation of pyrimidine dimers by adjacent thymidines.

192. (D) Researchers in 1928 worked with pathogenic (smooth) variants of *Streptococcus pneumoniae* that had been killed and thus were no longer capable of killing test mice, and living nonpathogenic variants (rough) of the same species, which also were incapable of killing mice. However, when these two nonlethal forms were mixed together and then injected into mice, the mice died. Later work showed that it was the DNA of the killed pathogenic form, and not the cell itself, that was capable of transforming the nonpathogenic living bacteria into pathogenic ones.

193. (D) The genetic code transcribed into mRNA is translated within ribosomes based on a three-base (triplet) code that indicates the corresponding amino acid that should be added next to the emerging protein. If one base is added or deleted from the template DNA, this causes the resulting reading frame to shift by one base in one direction or the other, thus the mutation is called a frameshift.

194. (B) In order for DNA replication to occur, the complementary base pairs within the double helix must be forced apart (melted) to allow the new base pairs to form between the template and newly synthesized DNA strand. Because this melting is accomplished by an enzyme, and because heat, like from the sun (*helios* in Greek), is known for melting things, this enzyme is called helicase.

195. (B) The nitrogenous bases used to make nucleic acids contain either a single ring structure (pyrimidines) or a double ring structure (purines; just remember that the bigger name goes with the smaller structure). Of the five bases used in biologic systems, three are pyrimidines (thymine, cytosine, and uracil) and two are purines (adenine and guanine). Just remember that AGs (as in Aggies) are PURe.

196. (A) Actually, two isotopes were used in these experiments, and they were carefully selected to label proteins (which contain sulfur but not phosphorus) and nucleic acids (which contain phosphorus but not sulfur). Viruses replicated in the presence of radioactive sulfur form radioactive protein capsids, while those replicated in the presence of radioactive phosphorus form radioactive nucleic acid genomes. When infecting new host cells, it was found that proteins from the capsid remained on the outside of the cells and thus were not involved in new virus formation.

197. (C) During the lagging strand synthesis of DNA replication, the discontinuous process necessitated by the 5′-3′ orientation of the template strand produces Okazaki fragments. These segments are connected by the continuous sugar-phosphate backbone of the template strand but have gaps in the backbone of the newly synthesized copy. These gaps are joined, or ligated, by the enzyme ligase.

198. (B) A mutation is defined as any change in the original DNA base sequence. Some of these changes may not produce any change in phenotype, and some may eventually produce a change in a single amino acid when translation occurs. Some, however, can involve extensive segments of DNA such as can be seen in gene duplication and, as here, an inversion of the original sequence of genes.

199. (B) Because of restricted base pairing seen in nucleic acids, guanine (G) always pairs with cytosine (C) on the complementary strand and adenine (A) always pairs with thymine (T). This means that the percentage of C in DNA will always equal the percentage of G, and the same holds true for A and T content. If some DNA contains 28% A, then the T content is the same (another 28%), and the total A-T content is 56%. This leaves 44% for the amount of G-C content, of which half is C, which therefore totals 22%.

200. (A) In order for DNA replication to occur prior to cellular division, the helix must be unwound and the complementary strands, held together by hydrogen bonding between the base pairs, must be separated (or melted). This process of unwinding and melting is accomplished by the same enzyme within the replisome complex known as helicase.

201. (E) The central dogma of biology is the flow of genotype coding to phenotype expression. DNA is replicated to produce accurate copies of the genome code; the genome code is then transcribed into RNA with instructions for the final translation process into protein. The DNA replication process is accomplished with a multienzyme and supporting protein structure known as the replisome.

202. (B) When the DNA double helix is unwound prior to replication, a tremendous amount of torsional stress is placed on the molecule, which if uncontrolled could produce a supercoiled tangle. To reduce this stress, one strand of the double helix is nicked as the replisome moves along the template. The enzyme responsible for this is gyrase, a form of topoisomerase.

203. (E) All living cells follow the same process of faithfully replicating their DNA genome, using that genome to produce RNA transcripts of the stored genomic code. This code, in turn, is used to produce the proteins required for all life functions, including the replication and transcription processes. This essential process is known as the central dogma. Retroviruses, however, violate this process through reversing the transcription process by using an RNA genome to form a DNA transcript, which is then used to continue the replication process. This can be summarized with the shorthand RNA → DNA → RNA → protein.

146 › Answers

204. (C) The bacterial genome is double-stranded DNA in the form of a circle. The process of replication during conjugation is unusual in that one template strand is nicked and peeled off as it is replicated and passed through the conjugation tube. This results in the circular template rotating, much as when sheets are pulled from a roll of paper towels, which is why this is known as rolling circle replication.

205. (C) Over 50 enzymes are associated with DNA repair. Some function to restore an incorrect single base substituted erroneously during DNA replication, while others are required to restore some fidelity following major structural trauma to the genome. When other less-radical fixes have failed, the cell must resort to SOS repair to recover as much of the original genome as possible.

206. (E) The electromagnetic spectrum extends from the very low energy and long wavelength forms of radio waves, up through the more energetic microwaves, and into the infrared and visible portions. Up through visible light, the photons do not contain enough energy to knock electrons from atoms and are thus known as nonionizing. Once into the ultraviolet wavelengths and on into X-rays, gamma rays, and cosmic rays, the energy content becomes ionizing. DNA can be damaged by these photons of shorter wavelength but not by those of lesser energy.

207. (C) Over the past few decades it has become understood that DNA is not as fixed in structure as once thought. In fact, it appears that massive rearrangements instigated by mobile genetic elements such as insertion sequences and transposons may be responsible for gene duplication and formation of new gene function. At the same time the cell requires that the genome remain as intact as possible. One of the genes used for DNA repair is *recA*.

208. (A) Both the DNA replication and RNA transcription processes are restricted in the way in which they move along a template strand and can read in only one direction. This prevents the respective processes from moving back and forth and manufacturing irreproducible copies and transcripts. Both of these processes can read the DNA template strand only from the 3′ to the 5′ end, and both involve synthesis of the new nucleic acid strand from the 5′ to the 3′ end.

209. (A) The replication process involves reading the base sequences of the DNA template strand and then synthesizing the DNA copy containing the complementary sequences. The enzyme that does this is a DNA polymerase. However, because retroviruses have distinctly different enzymes that also produce DNA synthesis but from an RNA template instead, the DNA polymerases must be distinguished from each other by additional descriptors. Thus the replication enzyme is more correctly known as DNA-dependent (reads DNA) DNA polymerase (makes DNA).

210. (C) The replisome complex is responsible for all of the functions necessary for the accurate replication of template DNA. This involves the unwinding of the double helix by topoisomerase, the actual copying of the template by DNA polymerase, the formation of short RNA primers for lagging strand synthesis by RNA primase, and the rejoining of Okazaki fragments by DNA ligase. ATP synthase, however, is required for ATP synthesis in chloroplasts and mitochondria.

Chapter 11: RNA Transcription

211. (E) Point mutations may or may not produce a change in the resulting amino acid sequence following translation, which may or may not change the function of the protein. Point mutations can be produced by a substance that modifies the nucleotide base on either strand of the DNA genome.

212. (D) In order for eukaryotic genes to be expressed and translated into proteins within prokaryotic cells, as is used in biotechnology applications, the noncoding segments that result in the exon regions of hnRNA must be removed because prokaryotes are not capable of posttranscriptional modifications to produce the mRNA. In the lab, already modified eukaryotic mRNA can be removed and the effective DNA gene can be reconstructed with the use of reverse transcriptase. This shortened version of the original DNA is known as complementary DNA, or cDNA.

213. (B) Any enzyme that reads a template nucleic acid strand to produce a complementary strand is called a polymerase. Normal transcription converts the codes in the form of DNA into an RNA transcript. However, because of retroviruses and their reverse transcriptase enzyme, the normal cellular transcription enzyme is most correctly called DNA-dependent (reads DNA) RNA polymerase (makes RNA).

214. (C) Under the central dogma of biology, DNA codes are converted to an RNA transcript that is used to produce a protein translation, summarized as DNA to RNA to protein. Reverse transcriptase is an enzyme that reverses part of this process by converting an RNA code to a DNA transcript. This reverse transcriptase enzyme is thus an RNA-dependent DNA polymerase.

215. (A) The expression of genes encoded in DNA is tightly controlled, as a cell does not wish to waste energy producing proteins that are either unnecessary at best or detrimental at worst. In prokaryotes this is simply done by the presence or absence of an active repressor that is a DNA-binding protein. However, the binding location, unique for each repressor, is always within a region of the operon downstream of the promoter region known as the operator region.

216. (D) The identification of the RNA polymerase binding region, the promoter, of the bacterial operon was discovered by a series of DNA blocking experiments where RNA polymerase was allowed to attach to DNA and the exposed portion of the DNA was then enzymatically degraded. When the RNA polymerase was removed, the intact DNA was sequenced and a highly conserved (i.e., unchangeable) series of bases was identified. This six-base sequence is known as the Pribnow box.

217. (B) The regulation of gene expression is very efficient. For example, if a cell finds itself in an environment where the temperature is too high for normal cellular enzymes to function, it is capable of repressing unnecessary transcription and simultaneously inducing a series of genes necessary for survival. This is accomplished through a process similar to the use of master and submaster keys. In the cellular world, the production of one sigma (σ) subunit, which normally promotes normal operations, is substituted with another, which promotes a different series of emergency genes.

218. (E) The bacterial operon consists of the basic components: the promoter region, the operator region, and a series of structural genes in sequence. Each of these regions are downstream from the previous segment. RNA polymerase always attaches to a highly conserved region found within the promoter region, thus the promoter region is the RNA polymerase binding site.

219. (D) In eukaryotes, transcription produces a form of RNA that contains both introns and exons, and the introns have to be removed before the strand officially becomes mRNA. This premature form is known as heterogeneous nuclear RNA, or hnRNA. Prokaryotes, which lack introns, thus do not need the posttranscriptional modification processes of eukaryotes; when they transcribe RNA with the required codes for producing a protein, it is already mRNA and available for translation.

220. (E) Eukaryotic RNA that contains the codes necessary for the production of proteins by ribosomes within the cytosol must be modified in form before it is allowed to leave the nucleus through the nuclear pores. Three things must be done before this can occur: first, the removal of the introns; second, the attachment of a 7-methylguanosine cap on the 5′ end of the molecule; and third, the addition of a series of adenine nucleotides at the 3′ end of the molecule known as the poly-A tail.

221. (E) Replication is the process of using a DNA-dependent DNA polymerase to manufacture a complementary DNA copy of a DNA template. Transcription is much the same except the nucleic acids are not the same form, so transcription either converts DNA codes to an RNA form or converts RNA codes to a DNA form. Normal transcription, as expressed in the central dogma, is thus converting a DNA template into a complementary RNA transcript.

222. (D) The pyrimidines cytosine and guanine are found in DNA, while uracil is normally found only in RNA. Thymidine, one of the two purines, is found only in DNA, while the other, adenine, is found in both forms of nucleic acid. Thus a segment of nucleic acid containing both purines and pyrimidines could be DNA, thus double stranded, or RNA, which is single stranded and contains ribose in its sugar-phosphate backbone.

223. (B) An enzyme identified as an RNA polymerase synthesizes RNA from a nucleic acid template. In normal cellular processes, this template is always DNA, and the common enzyme is a DNA-dependent RNA polymerase. However, some RNA viruses carry inside their capsids at least one copy of a viral enzyme that permits the copying of their RNA genome into a complementary RNA copy, which then serves as a template for replicating their genome. This enzyme is thus an RNA-dependent RNA polymerase. In either case, RNA is the product.

224. (B) The promoter region of the bacterial operon is the location where the RNA polymerase binds to begin transcription. To do so, there can be no repressor bound to the DNA within the operator region, otherwise transcription of the genes downstream of the operator cannot occur. A mutation within the highly conserved sequence of the promoter could only prevent the binding of the RNA polymerase and thus permanently prevent transcription of that operon.

225. (A) Eukaryotic RNA, when transcribed for messenger function, must undergo three posttranscriptional processes before the resulting mature mRNA can leave the nucleus. The synthesis of bacterial mRNA begins with a molecule of ATP at the 5′ end, and because it lacks introns, it can be immediately translated as soon as ribosomes can attach. In fact, in bacteria, transcription and translation of a single message can occur simultaneously.

226. (B) Three- or four-letter codes are used to aid in the identification of specific genes and their respective functions or products. The code *trp* identifies the repressible operon associated with the manufacture of the amino acid tryptophan used in protein synthesis, which is a continuous process under growth conditions. It is repressible because under unfavorable growth conditions protein synthesis ceases, as does the requirement for the synthesis of the amino acid.

227. (E) Prokaryotic mRNA requires no modifications prior to its use for translation. However, the genomic structure of eukaryotes is much broader and complex. In order to keep newly synthesized but nonsensical hnRNA from going into the cytosol for translation, it must be restructured and made into a mature form. This includes the removal of introns and the rejoining of the coding exons, and the attachment of a protective 7-methylguanosine cap on the 5′ end and a polyadenylated tail on the 3′ end.

228. (E) Thymidine is a pyrimidine nucleotide found in DNA. Its functional equivalent, as both base pair with adenine, is uracil. Any long segment of a nucleic acid lacking thymidine would probably be RNA, which exists primarily in a single-stranded form and whose sugar-phosphate backbone contains ribose.

229. (A) Bacterial operons are classified as repressible, inducible, and constitutive. Repressible operons are those that are normally expressed but, when not needed, can be shut down (repressed). Inducible operons are the opposite in that they are normally repressed but can be derepressed (induced) when needed. Constitutive operons are neither repressible nor inducible, and are continually expressed, as they are commonly associated with vital metabolic processes.

230. (A) DNA blocking studies identified the regions on the bacterial genome that were consistently found as indicating the binding location of RNA polymerase. While this binding site contained areas of high base variability, one segment with a highly conserved six-base sequence was always present: TATAAT. This conserved sequence is also known as the Pribnow box. The eukaryotic equivalent sequence is TATA.

231. (E) If there is any change within the highly conserved TATAAT sequence, then RNA polymerase will not bind, and as a result, that operon will not be expressed (it will be silenced) until the original sequence is restored. Pending that and assuming the expression of the operon is vital to the survival of the cell, the mutation would be lethal.

Chapter 12: Translation

232. (C) The components of ribosomes, which are composed of two different sized subunits, are identified based on their sedimentation coefficient or svedberg (S) value. Eukaryotic ribosomes are larger and more complex than the ones in prokaryotic cells. Those found in the former are identified as intact 80S ribosomes, while the latter are 70S.

233. (E) An active ribosome will consist of not only the ribosome itself but also the additional two forms of RNA required for translation: mRNA, which contains the code that will be translated, and tRNAs, which deliver the appropriate amino acid in the proper sequence. The ribosomes that are required for protein synthesis are composed of both proteins of assorted molecular weights and several strands of rRNA.

234. (E) The assembly of the ribosome with its two subunits, along with the mRNA to be translated and the assortment of amino-acid-charged tRNAs, is spontaneous. Even the movement of the ribosome along the mRNA requires no energy expense. However, the formation of the peptide bond formed between the elongating peptide and the new amino acid being added requires the expense of ATP.

235. (C) Three sites have been identified that have specific activities within the translating ribosome. The first, the E site, is the location where a tRNA that has donated its previously attached amino acid is about to exit the ribosome. The second, the P site, is where the elongating protein is located following the addition of the most recent amino acid. The third, the A site, is where the next charged tRNA, with its corresponding amino acid, aligns its anticodon sequence with the codon triplet on the mRNA.

236. (E) DNA polymerases are required for the synthesis of DNA regardless of the template strand read. The same is true for RNA polymerases that synthesize RNA. Translation of the genetic code into functioning proteins involves no enzymes, just the active ribosome.

237. (A) If the DNA codes ATG-CGT, it is transcribed into its complementary RNA sequence of UAC-GCA (remembering that there is no thymidine in RNA and that uracil is in its place). This mRNA sequence represents the codons within the genetic code. As each tRNA aligns with the mRNA in the ribosome, their complementary sequences present in the second hairpin loop, correspond to the codons, and are thus known as anticodons. The anticodon for UAC would be AUG and the anticodon for GCA would be CGU.

238. (D) Peptide synthesis requires that all components associated with the process be present. This not only includes the proteins and rRNAs of the ribosome itself, but also the coding mRNA and the tRNAs that shuttle in the proper amino acids.

239. (A) In order for protein synthesis to occur, there has to be an interface between the world of amino acids and proteins and the world of nucleic acids and genetic code. The molecule that contains both connected by a covalent bond of much greater strength than the hydrogen bonds that bring nucleic acid strands and proteins into close association is a tRNA molecule that has been charged with its appropriate unique amino acid.

240. (B) There are 20 essential amino acids encoded within the genetic code, but there are many more possible codon combinations. Besides the redundancy, where more than one codon can be used for any single amino acid, there are four special codons. Three of these, the stop codons of UAA, UAG, and UGA, have no corresponding amino acids and, when present, cause the conclusion of synthesis and release of the new peptide. The fourth, AUG, codes for the first amino acid of every protein, methionine (f-methionine in bacteria).

241. (D) The *lac* operon is the archetype inducible operon that is normally repressed in the absence of lactose but can be induced in the presence of lactose and under starvation conditions. β-galactosidase is one of the three enzymes produced for the use of lactose when this operon is induced and is also commonly used as a reporter gene in genetic transformation studies.

242. (D) Protein synthesis can be regulated with two basic mechanisms: the breakdown of the mRNA that codes for a respective protein after it has been transcribed and the prevention of the synthesis of that mRNA. Proteins that bind to the operator region of bacterial operons, thus preventing transcription of mRNA and the eventual translation of it into a peptide, are known as repressors.

243. (B) Translation occurs in three phases. The first stage, initiation, is when the inactive ribosome opens and the mRNA aligns with segments of rRNA and then reassembles in preparation for translation. The third, termination, occurs when a stop codon is encountered in the mRNA message, resulting in the disassociation of the ribosome and the completed peptide. The middle segment, which consumes the bulk of the translation process, is known as elongation, as it is during this period that the peptide is lengthened, amino acid by amino acid.

244. (B) Both prokaryotic and eukaryotic ribosomes consist of a larger and a smaller subunit. In the case of the former, they are identified as 50S and 30S, while in the latter, they are 60S and 40S. While differing in specific components, all ribosomal subunits consist of both proteins and rRNA.

245. (C) In order for translation to occur, all of the molecular components must be simultaneously present: mRNA containing the required code, tRNA to conduct the appropriate amino acid to the A site, rRNAs to provide alignment, and various proteins to provide assorted structure and function. DNA is not associated with translation, only replication and transcription.

246. (A) In order to assemble a ribosome, mRNA is transcribed and escorted out of the nucleus and into the cytosol for the synthesis of ribosomal proteins. These proteins are then escorted back into the nucleus. The presence of newly synthesized rRNA initiates the assembly with the proteins to produce a ribosomal subunit. This subunit is then escorted back out of the nucleus where subunits can self-associate for translation.

247. (C) When considering just the numbers of the units available, there are only three strands of rRNA (four in the case of eukaryotes), one or two loose tRNAs and their associated amino acids, but a total of at least 55 ribosomal proteins in prokaryotes and more than 82 for eukaryotes.

248. (E) Codons consist of three bases in sequence. At each of those positions there can be any one of the four RNA bases (A, U, G, or C). Since the total number of possible combinations must consider the possibility of any base at any position, there are 4 possibilities at the first position times 4 possibilities at the second position times 4 possibilities at the third position, or 4^3. This means there are 64 possible codons coding for the 20 essential amino acids.

249. (B) The termination phase of translation occurs when one of the stop codons is encountered in the A site of the ribosome. The stop codons have no corresponding charged tRNA escorting in an associated amino acid, and because the process cannot continue, the ribosome disassociates and the protein is released. Of the 64 possible codons, three are stop codons.

250. (A) Whenever a carbon atom is connected to any other atom of carbon or hydrogen, energy is stored in the associated covalent bond. Generally, the bigger and more massive an organic compound is, the greater the energy content. In order to store this energy, it must be transferred from one energized molecule to another during synthesis. In the case of translation, energy from a molecule of ATP is transferred to the emerging peptide to form a peptide bond between amino acids.

251. (B) When RNA polymerase begins transcription, the initial DNA bases transcribed are not within the portion of the operon that is used to code for any portion of the resulting protein but rather lie within the operator region. As the RNA polymerase continues synthesis, it eventually encounters the DNA sequence TAC. This transcribes into the start codon AUG on the mRNA and signals the initiation point of the eventual protein synthesis. All of the RNA prior to this start signal, which is lying upstream, is considered the untranslated leader sequence. All RNA bases downstream from the start codon were transcribed only from the exons left after posttranscriptional modifications occurred within the nucleus.

252. (E) Once a strand of mRNA enters the cytosol of a cell, a specific series of bases within the noncoding leader segment known as the Shine-Delgarno sequence signals the location of ribosomal assembly around the mRNA. Once assembled, the ribosome moves down the mRNA codon by codon, with each signaling the addition of another amino acid. Once it has moved down and the Shine-Delgarno sequence is again exposed, another ribosome can assemble and begin translation of the same message, and this process can repeat over and over. The resulting cluster of ribosomes is known as a polysome.

Chapter 13: Genetic Engineering

253. (C) Research into the possible use of viruses as antibacterial agents uncovered a bacterial defense mechanism that protected the bacteria from viral invasion. Researchers found that when the viral DNA entered the bacterial cytosol, it was quickly fragmented by bacterial enzymes known as nucleases. Since it was found that the enzymes actually cut the DNA at specific, or restricted, sites, they became known as restriction endonucleases.

254. (E) The polymerase chain reaction (PCR) is used to amplify the quantity of specific DNA in the laboratory using a process known as thermal cycling. In order for the DNA polymerase to initiate replication, it must associate around a short segment of double-stranded DNA. This double-stranded region is formed when a short 18–25 base computer-designed DNA primer is allowed to hybridize very specific sequence in the target DNA, one for both strands present. While more than one pair of primers can be used, multiplexed reactions involving more than one pair become much less efficient.

255. (A) Genetic manipulation formerly consisted solely of random mutagenesis and massive screenings of mutants. Newer techniques, based extensively upon restriction nucleases, use mobile genetic elements such as transposons, phages, and plasmid constructs. The letter *p* identifies a plasmid (or protein in some cases); pBR322 is a plasmid of known genetic content commonly used for the insertion of desired genes into bacteria.

256. (C) PCR requires the use of a DNA polymerase that is thermal-stabile because the technique elevates the reaction temperature to over 95°C in order to melt, or separate, the two complementary strands of target DNA. At this temperature most enzymes are very rapidly denatured and lose their effectiveness. *Taq* is shorthand for the DNA polymerase originally recovered from the thermophilic bacterium ***T****hermus **aq**uaticus* whose enzyme can function for hours at 95°C.

257. (D) Once DNA has been manipulated and constructed by a researcher, it must be inserted into a cell so the modified DNA can be transcribed and translated into the desired proteins. One such technique involves very small latex microbeads coated with the desired DNA. A device similar to a very small shotgun, known as a gene gun, can deliver hundreds of these microbeads into an entire cell sheet or tissue at one time.

258. (E) There are a number of techniques available to get DNA past a membrane and into the cytosol of bacteria or nucleus of eukaryotic cells. This can be done one cell at a time by microinjection with a very thin glass needle, or more randomly in the lab by techniques such as electroporation, general or specialized transduction, or transformation.

259. (C) In order to identify which genes are being expressed in a cell under certain conditions, the cell is disrupted and the mRNA is extracted. These mRNAs represent every gene that is being transcribed at the time of extraction. In order to identify which genes are represented, researchers will separate the RNA species by electrophoresis and then blot, or transfer, the RNA from the electrophoresis gel onto a material such as a thin sheet of nitrocellulose. By using specific reporter-labeled DNA or RNA probes, the desired band can be identified. This technique is known as a northern blot.

260. (A) When probing for specific nucleic acid or protein sequences, a reporter molecule must be attached to a small molecular probe that will attach to the desired sequences only. These reporter molecules are commonly radioactive, and their position on a blot is detected by a layer of X-ray film. The resulting image is known as an autoradiograph. This can be used for the detection of specific proteins in western blots, specific RNA in northern blots, or specific DNA in Southern blots.

261. (B) Short segments of either DNA or RNA are often referred to as oligonucleotides, where *oligo-* comes from the Greek for "short or few." They can easily be labeled with a reporter molecule for the identification of complementary nucleotide sequences. These reporter molecules may be radioactive materials, enzymes, or chemiluminescent.

262. (D) The separation of molecules is a very common analytical technique and can be done by fractionation, distillation, chromatography, antibody binding, gradient centrifugation, or electrophoresis. This last technique involves the movement (*-pheresis*), thus separation, of molecules by electricity (*electro-*) and is commonly used to separate fragments within a polysaccharide such as agarose prior to identification by subsequent blotting techniques.

263. (B) Anywhere from 250,000 to 500,000 children become blind annually in third-world countries because of a shortage of vitamin A in their diet. The addition of small amounts of a vitamin A precursor such as β-carotene could prevent this. Researchers have inserted the genes necessary for the production of β-carotene into rice, which normally is vitamin deficient but is a major staple for over three billion people. The presence of the β-carotene is indicated by the light orange or golden color of the grain of this genetically modified organism (GMO).

264. (C) DNA commonly travels from bacterial species to bacterial species horizontally by a number of natural mechanisms. This may be by the acquisition of naked DNA from a dead organism (transformation), or the movement of DNA from host to host by viruses (transduction and infection), or by sigma mode DNA replication of genomic contents by bacterial conjugation. Electroporation is a lab technique that greatly increases the effectiveness of bacterial transformation.

265. (D) There are a couple of characteristics of restriction endonucleases that make them very valuable for research due to the way they protect bacteria from viral infection. First, they cut double-stranded DNA at very specific sequences. Second, these cuts are not usually straight across the molecule but rather in a jagged fashion that produces short, overhanging, single-stranded regions that easily reconnect to the same complementary sequences regardless of their original source. These easily reconnectable regions are known as sticky ends.

266. (A) One of the key tools used for the identification of specific genes involves the electrophoretic separation of molecular fragments followed by capillary transfer onto a thin blotting material and the use of labeled probes. When the substance involved is DNA and the probes are oligonucleotides, the technique is known as the Southern blot.

267. (A) Electrophoresis uses electromotive force to drive molecules through a molecular sieve, which causes the separation of these molecules based on some inherent characteristic. These techniques may separate based on size, charge, mass, or length, with the material selected for separation chosen because of the characteristic used. While agarose is most commonly used for the separation of oligonucleotides, polyacrylamide is most commonly used for the separation of proteins.

268. (A) The term *genetically modified* implies an artificial or lab-manipulated organism. While the DNA from various species has been detected in the cells of protozoans and humans, the use of the gene for human insulin required the laboratory conversion of mRNA into cDNA before insertion into bacteria resulted in the commercial production of insulin.

269. (B) We are still incapable of detecting only one strand of nucleic acid. However, the detection of thousands or millions of strands is a much simpler and practical matter. The most common technique of amplifying one to millions for copies of DNA involves the use of DNA polymerase and thermal cycling, which results in the doubling of the amount of DNA after every cycle. This now very standard technique, first developed in 1986, is known as the polymerase chain reaction, or PCR.

270. (E) RFLP stands for a DNA fragment comparison technique known as restriction fragment length polymorphism. Nucleic acid from various organisms is subjected to restriction endonuclease digestion, which results in the formation of tens to perhaps hundreds of fragments. By comparing the fragments from different organisms following electrophoresis, the patterns can be matched for identification purposes.

271. (C) Blotting is a very common and useful technique and can be used for a variety of purposes and with a variety of materials. The electrophoretic separation followed by probe identification of specific DNA sequences is known as the Southern blot. When done with RNA, it is known as a northern blot. If performed with a protein, it is called a western blot.

272. (D) After performing genetic manipulation, a researcher needs to know if the process was successful. In order to confirm this, some mechanism is added to the process that can report success. Commonly, for transformation studies, the reporter gene placed in a plasmid might indicate a successful transformation by changing the organism from sensitive to resistant to a certain antibiotic or by the conversion of a dye in the growth medium that changes the bacterial colony from white to blue.

273. (A) Successful organisms have numerous protective genes. Some of these genes found in bacteria protect them from specific antibiotic attacks, and this protection is known as resistance. Some common plasmids contain numerous resistance genes that can move horizontally from species to species, spreading their resistance capabilities. These plasmids are known as R (for resistance) plasmids for that reason.

Chapter 14: Origin of Life

274. (C) When trying to reconstruct the atmosphere thought to have existed shortly after the formation of the earth, researchers have relied on chemical processes derived from entirely inorganic materials. This is because the atmosphere must have formed before the advent of life. Since oxygen is highly reactive, it is usually not found in inorganic conditions unless it is in combination with other elements. The early atmosphere would have been devoid of oxygen gas because atmospheric oxygen is of photosynthetic origin.

156 › Answers

275. (D) It is thought that the earth formed about 4.5 billion years ago. The earliest evidence of life comes in the presence of stromatolites, thought to have been built through a mechanism enhanced by cyanobacteria, which date back to 3.5 billion years ago.

276. (A) Molecular oxygen is extremely reactive, and it appears that life processes could not have originated within an atmosphere filled with it. While silicates may have contributed to early polymerization reactions, the most important compound in biologic reactions is water.

277. (C) DNA, regardless of what it might code for (rDNA contains the genes required for ribosomal function) or its source (mtDNA is that which is found within mitochondria) is a rather complex molecule. It is thought that the earliest nucleic acid formed abiotically was most likely basic RNA because of its relative simplicity.

278. (E) Fossils can be dated based on the presence of radioactive elements they contain. Since every element has radioactive isotopes, and since the decay rate for any given isotope is constant and unaffected by outside influences, the ratio of isotopes in a sample can be used to calculate the age of that sample. The initial ratio has to be assumed in these cases.

279. (D) Proteins are composed of their monomeric subunits of amino acids connected through peptide bonds. In cells these amino acids are assembled within a ribosome that could not have been present in the early atmosphere. Proteins cannot be constructed abiotically *de novo* without the presence of previously existing amino acids.

280. (E) Proteins are polymers, and the construction of a polymer, which is a structure composed of repeatedly added monomers, requires some mechanisms that provide regularity for the repetitive process. Without the regularity of action inside a ribosome, it is thought that repetitive crystalline structure within some clays provides the best potential mechanism for the abiotic construction of proteins.

281. (C) Given that life had to have an origin, early researchers thought to envision the early atmospheric conditions that might have led to the abiotic formation of organic compounds and polymers. John Haldane and Aleksandr Oparin, both in the 1920s, presented their ideas as to the contents of this early reducing atmosphere.

282. (C) The simplest of cells are prokaryotes, which lack a nucleus and any internal organization resembling an organelle. Eukaryotic organisms have double membrane–bound structures within their cytosols in the form of the nucleus, mitochondria, and, in plants, chloroplasts. It is thought that these more complex cells were initially formed when some bacteria took up symbiotic relationships with them about 1.5 billion years ago.

283. (B) The endosymbiotic theory posits that the cellular organelles we now call mitochondria and chloroplasts were once commensals that eventually became symbionts within a host cell. This is supported by the bacterial-like ribosomes found in these organelles and the presence within these structures of their own DNA, some of which might have migrated to the nucleus. What this theory does not adequately explain is the nuclear membrane's triple layered structure filled with pores.

284. (D) Sedimentary rock, as it is found around the world, is observed in five different basic layers, representing the major eras of the geologic record. Each of these eras is further divided into periods, which are further subdivided into epochs, based on their fossil contents. The five major eras are Archaean, Precambrian, Paleozoic, Mesozoic, and Cenozoic.

285. (E) Haldane and Oparin first envisioned the contents of the primitive atmosphere that must have been present for life to begin abiotically. These ideas led to the research of Miller and Fox, who contained these gases within a globe and used the energy of electrical discharges (to simulate lightning), which then led to the production of some simple organic compounds.

286. (A) In order for early fossils to be analyzed, they first must have been formed under the right conditions, been driven to the surface by tectonic action, and survived being exposed to the elements up until discovery. One of the best locations where these conditions have all been met for Precambrian fossils is in southern Australia.

287. (E) The five major geologic eras are closely linked to the predominant life-forms found as fossils within their rocks. The rocks of these eras include evidence of bacteria only (Archaean), early invertebrates (Precambrian), many varieties of invertebrates and land plants and insects (Paleozoic), dinosaurs (Mesozoic), and mammals (Cenozoic).

288. (E) The most recent geologic era includes the older Tertiary period (including the Paleocene, Eocene, Oligocene, Miocene, and Pliocene epochs) and the newer Quaternary period with the Pleistocene and Recent epochs (in which we now reside). These all fall within the most recent Cenozoic era.

289. (A) If we presented the eras as the time on a 12-hour clock with now being noon, the current Cenozoic era would have begun just three minutes ago (equivalent to 65 million years ago), the Mesozoic era would have started 51 minutes ago (about 250 million years ago), and the Paleozoic era would have started just one hour and 56 minutes ago at just after 10:00 A.M. (or 570 million years ago).

290. (B) There are over 35 animal phyla that span all the way back in the fossil record to the Precambrian era, which lasted over one billion years. The Cambrian period immediately followed, from about 570 to 505 million years ago. Fossils linked to that period include all of the major animal phyla except Chordata, which finally appears in the immediately subsequent Ordovician period.

291. (D) When compared to the other rock-based planets of the inner solar system, the earth is unique due to its magnetic field, which is generated by its molten core. This molten core is also thought to have produced another unique characteristic, that of continental drift. Based on fossil evidence, as well as on current measurable rates of drift, it is thought that the seven continents we can see now were once combined into two major land masses about 135 million years ago called Gondwana and Laurasia, and once even a single land mass about 200 million years ago called Pangea.

292. (B) Humans have an innate need to identify things by giving them names. This holds especially true in the realm of biologic taxonomy. The broadest traditional category, which contains the greatest number of species, is the kingdom. Below that, in order of greater specificity and less inclusion, are division (or phylum), class, order, family, genus, and finally, species.

293. (C) Within the most recent Quaternary period, we find evidence for wide variations in mean planetary temperatures, with ice ages intermingled with global warming. The glaciers of these ice ages covered most of North America down to the Missouri and Ohio rivers and bound up enough water to drop the mean sea level by up to 100 meters.

294. (D) Several places within the fossil record give evidence of massive and sudden extinctions of many life-forms. The presence of the element iridium, most likely of meteoric origin, is found in rock layers between Mesozoic and Cenozoic sediments at the end of the Cretaceous period, which also corresponds to the disappearance of most dinosaurs from the fossil record.

Chapter 15: Evolutionary Mechanisms and Speciation

295. (A) When observing fossils of the heads of dinosaurs, there are some structural differences that can be noted when comparing them to mammals. The first is that in dinosaurs there is only one bone connected to the eardrum for transmitting sounds. Second, there are a couple of extra bones within both the lower and upper jaws. It appears likely that the ossicles of the inner ear of mammals were derived from the relocation and refunctioning of these missing dinosaur bones.

296. (A) When the word *mutation* is mentioned to a layman, degenerate slime remnants of previously normal organisms are most often brought to mind. Even in the sciences it is easy to confuse terms that describe a mutation in the genotype with a mutation in a phenotype. Most correctly, a mutation is any change within the original DNA sequences of an organism.

297. (D) Not all dinosaurs died off at the end of the Cretaceous period 65 million years ago. Large numbers of animals still exist that possess specific bone structure similarities to the archosaurs of that period. These animals, all of which excrete uric acid rather than the urea of mammals, include crocodilians and birds.

298. (E) When individuals of one species migrate away from or become separated from their kin, they commonly begin to inhabit a slightly different habitat. Over the subsequent generations, as small changes in surviving progeny increase the survival of specific phenotypes, physical differences appear when compared to the original. These changes are called adaptations, and where new variations spread out from the original, the process is called adaptive radiation. This process is thought to explain the many variations of Galapagos finches.

299. (B) The phrase "survival of the fittest" is commonly bantered about. But what does that really mean? The strongest? The biggest? The fastest? The one that can hide the best? While all of these can be measured, they do not necessarily indicate what is the fittest in evolutionary terms. The best measure is the type or strain that stays around the longest, and that usually means that they have more surviving progeny than any other. The best measure of effective evolution is reproductive success.

300.(C) Several lines of evidence suggest the origin of bird feathers. The first is that they form a protective cover that also serves a function in attracting mates and reproductive success. Another is that the feather's structure and means of insertion are most closely related to the presumptive ancestor group of archosaurs, and that feathers have been observed within the fossil record with them. Thus it appears that feathers are derived from the scales of reptiles.

301. (C) Groups of organisms from common ancestors can become isolated from each other, thus leading to speciation, by different mechanisms. One subgroup might start preferring a different mating ritual, thus separating behaviorally. One subgroup might develop a change in reproductive structure that separates mechanically. But when identifying the change in habitat, this indicates an ecological isolation.

302. (A) Just because an organism appears within the fossil record does not mean that it is extinct. Some designs or adaptations have been exceptionally successful and enduring, while others have been selected against. The mosses are one of the oldest successful land plants and precede all humans and current flowering plants, reptiles, and even fish species.

303. (E) All of the organisms listed are mammals, so the question becomes, "Which was the first to appear in the fossil record?" The earliest mammals appear as a group known as therapsids, and they coexisted with dinosaurs through the Mesozoic era, which proved to be a superior adaptation following the Cretaceous extinction. These early therapsids most closely resemble current-day shrews.

304. (B) Charles Darwin visited the Galapagos Islands in 1831. While there he made a series of observations where he compared the sizes and shapes of the beaks on finches in relationship to their location and available food sources. When he assumed that they were all descendants from a common ancestor, he was led to conclude that each specific adaptation was the result of an evolutionary mechanism that led to speciation.

305. (D) Up until fairly recently, the most commonly accepted explanation for the disappearance of dinosaurs from the fossil record was either that they were outcompeted by the later dominant mammals or that an extensive global calamity, such as a decades- or century-long drought, eliminated the largest land animals. However, the discovery of a very large crater just off the Yucatàn Peninsula, along with the findings of the meteorite-associated element iridium in a specific sedimentary layer worldwide, both of which are dated to the same time period, leads to the current theory that the earth's collision with an asteroid was the cause.

306. (D) Many researchers feel that we are presently in the midst of a major period of extinction. One of the major human characteristics that has led to our success over all the planet is our ability to adapt the environment to our needs rather than forcing ourselves to adapt to the environment. It is thought that we have changed too much too quickly, and as a result too many species have not been able to adapt quickly enough to survive.

307. (E) Mammals dominate the land, but fish dominate all waters. Insects, on the other hand, outnumber both combined. However, the most successful group, evidenced by the fact that they outnumber every other form of life, occupy every niche available on the planet, and have been here longer than any other, are bacteria.

308. (A) Birds have long been observed for their fascinating behaviors, and it is obvious that each species has unique manners that separate it from all others. These behaviors include feeding methods, mating displays, nesting preferences, and territorial choices. However, it is the specific songs that have been learned by fledglings from their parents, which show evidence of a form of "baby talk" and regional dialects, that most clearly produce species isolation.

309. (A) Phenotypic similarities do not necessarily indicate genetic similarities, as we see in cases of convergent evolution where environmental influences can have an effect. Additionally, just because two organisms appear similar, it may be a case of mimicry rather than phylogenetic similarity, and this would not lead to successful matings. In the case of fossil classification, however, all we have to go on is structural similarities.

310. (B) Mechanisms that have influence on whether or not fertile progeny will result from a mating can be prezygotic (meaning before the mating produces a zygote) or postzygotic (meaning that something within the zygote itself is involved). Prezygotic isolation can be behavioral (wrong mating ritual), physiologic (wrong chemical signals in the form of pheromones or seasonal variations in conception cycles), or even geographic (wrong continents). Hybridization issues, usually because of mismatched chromosomes, can also prevent progeny with the potential for fertility.

311. (D) Gene flow is the movement of alleles within a species. This may involve movement from one population to another in the case of immigration or emigration. Because the movement of alleles can only resort from mating, and because successful mating can only occur within the same species, this excludes gene flow within a genus (multiple species), community, or ecosystem (multiple genera).

312. (A) Dolphins and their kin are actually closely related to ungulates, and both are fairly recent on the evolutionary scene. While rodents do closely resemble the earliest forms of mammals, they too are newcomers. Echinoderms, being more primitive as nonplacental mammals, might be a good choice, but the earliest mammals, newly derived from therapsids, were very small, tree-dwelling, nocturnal, insect-eating animals.

313. (E) Evolutionary changes occur as a result of genetic changes regardless of what the older term *adaptation* might imply. Changes in DNA occur first and can then be passed on through sexual reproduction, randomized through crossing over during meiosis, and then dispersed through movement such as immigration.

314. (C) Alfred Wallace was a well-known biogeographer who did vast exploratory research in both Malaysia and Brazil, drew the same conclusions about natural selection as Darwin, and published his various findings extensively. He is considered the codiscoverer of evolutionary theory.

315. (B) On occasion, such as following a calamitous event, the number of individuals within a population is greatly reduced. This sudden reduction also greatly reduces the number of alleles possessed by the residual population, thus greatly reducing the genetic variety within a population and increasing its susceptibility to further disasters. This is known as the bottleneck effect and is the opposite of genetic drift.

Chapter 16: Viruses

316. (E) All cells are capable of metabolizing substances to drive cellular functions, but viruses must cannibalize these from their hosts, as viruses lack all metabolic abilities. All known viruses contain either DNA or RNA but not both, a characteristic belonging only to cells. All cells divide, either by mitosis or binary fission, when an original mother cell splits into two identical daughter cells. Viruses, on the other hand, replicate by an assembly-line-type process.

317. (B) Viroids are actually not viruses at all but rather are just naked strands of RNA that can infect cells and produce pathology. Multipartite viruses are similar to multigenomed viruses, such as influenza virus, in that their genome is divided into more than one segment. But in the case of the former, each genome segment is also packaged separately into its own independent capsid. The one thing they have in common is that they both are found only in plants and must be transmitted by the bite of feeding arthropods.

318. (B) All viruses are composed at a minimum of a nucleic acid genome and a protein coat called a capsid. Some viruses have very pronounced projections sticking out from the capsid that have been referred to as spikes. Normally these spikes act in a manner similar to the pili of bacteria and are instrumental in both attaching to and entering the host cell.

319. (E) The genome of ambisense viruses acts as both the sense and antisense template. This means that one part of the genome serves as the sense strand and thus can be immediately translated into viral proteins, while the other part serves as a template for the manufacture of a complementary copy, which then serves as an mRNA available for immediate translation. However, no known viral genome can be translated in both directions, as this is a limitation imposed by the host ribosomes, which can read in only one direction.

320. (C) All lipid bilayers in nature are associated with cellular membranes. But viruses do not have membranes, which serve cells by controlling the access of substances into the cytosol. Instead, when some viruses bud through a host cell upon maturation, they can rip out a section of that membrane much like a golf ball punching through a sheet of plastic food wrap. When this happens, the former cell membrane is called an envelope.

321. (A) Some viruses produce acute infections. This means that they have a generally short incubation period, then produce a rapid onset of characteristic symptoms, and finally are cleared from the host within a matter of days to weeks. On the other hand, some viruses infect a cell and then integrate their genome into that of the host cell and become dormant until some trigger reactivates the replication cycle of the virus. These are known as cryptic viruses, and they produce latent infections.

322. (B) The use of cell or tissue culture is common in viral propagation. Some cell lines have become immortalized either through a viral infection or carcinogenic process and are called transformed cell lines. Some cell lines thrive for 30 to 50 or more divisions, but then start to lose viability and die off. These are known as diploid cell lines. Some viruses will not propagate in either transformed or diploid cell lines, but instead require cells no more than 10 or so generations removed from the original *in vivo* source. These cell lines are referred to as primary cell lines.

323. (B) All viruses have a common set of characteristics that determine that they are, after all, viruses. These characteristics include that they lack metabolic pathways, that they vanish immediately after uncoating within the host cell only to reappear again at the end of this eclipse period, and that each virus is an exact copy of every other of its kind, a structural consistency that enables them to be crystallized. What all viruses lack is a cell membrane, a characteristic of all living cells.

324. (B) The protein coat that is possessed by all viruses is generally composed of a relatively few specific viral proteins often produced late in the infection cycle, although some pox viruses contain scores of the coat proteins. Each protein that is associated with the viral capsid is a capsid monomer, or capsomer.

325. (C) While *Chlamydia trachomatis* is an obligate intracellular parasite, it is a bacterium, not a virus, and requires a specific human host cell. Of the remaining, the rabies virus, the herpes simplex virus, and the orf virus are all viruses of mammals. Only the tiny φx174 is a bacteriophage.

326. (E) Viruses, because they are not living cells, do not follow all of the common rules associated with cells. All cells, for example, contain DNA that is only double stranded and RNA that is single stranded. Viral genomes, however, may be double-stranded RNA, partially double-stranded DNA (part double stranded and part single), or even single-stranded DNA. A parvovirus, in fact, possesses a genome in this last form.

327. (D) A virus is an obligate intracellular parasite simply because that it cannot replicate without using the cellular machinery already in place within a host cell. When infected with an acute virus, the proteins that the cell starts to synthesize as directed by the viral genome tend to shut down the pathways required for maintaining cellular homeostasis and instead are directed to manufacture only viral proteins and genomic material.

328. (D) Viruses can be found in seawater, ice, and soils, and even have been detected in fog. In fact, wherever cells are found, their viral parasites are found right alongside and usually outnumber their host cells by several orders of magnitude. Viruses of animals, plants, and bacteria are commonly studied.

329. (D) There is a well-known transformed cell line called HeLa cells. These cells were originally cultured from the cervical tumor that eventually ended the life of Henrietta Lacks in October 1951. These cells were transformed because they were infected with a common carcinogenic genotype of human papillomavirus, which is the primary cause of cervical cancer.

330. (E) While some viruses have envelopes or peplomers, are diploid or multigenomed, or contain enzymes or host cellular debris, all viruses have the common characteristic of a protein coat surrounding a single form of nucleic acid.

331. (C) Viruses are almost always the cause of some disease. In fact, the search for the etiologic agent of a disease is what most often leads to the discovery of a virus. However, many diseases are caused by other pathogens, including bacteria, fungi, protozoans, and animal parasites. Of those listed, only syphilis is not caused by a virus. Rather, it is caused by the bacterium *Treponema pallidum*.

332. (C) Viruses do not grow, develop, or divide; viruses replicate. They are assembled within a host cell much as a car is assembled in a factory. But before they begin their replication cycle, they must gain access to the host cell machinery by getting into the cell. This is accomplished by the attachment of viral proteins, acting as ligands, to specific host cell proteins, acting as receptors. Once attached, the virus gains entry through the cell membrane.

333. (E) Acquired immune deficiency syndrome (AIDS) is the final stage of a disease that is currently classified as a 100% fatal disease in that it can be treated just as diabetes can be treated but not cured. Human immunodeficiency virus, also known as the AIDS virus, infects a person and if untreated, eventually destroys the T-helper cells that are a key element in immunologic protection.

334. (B) Infectious agents differ from toxins in that they amplify in the host while toxins are diluted, meaning that the former can be passed host-to-host while the latter must come from a point source. Amplification, in every case but one, means that the agent must be either a virus or an organism. Prions, the one exception, are mutated versions of a normal neuron-associated protein that, once it gains entry, causes normal host proteins to cluster just like the mutant form. The clusters accumulate and cause spongiform encephalopathies.

335. (B) Viruses cause the death of a host cell by preventing the host cell from maintaining its own homeostasis. This is because the virus commandeers the host molecular machinery for its own replication purposes. The less the cell is able to care for itself, the greater the damage from the virus. In some cases, relatively slow production allows the cell to survive with minimal effects. However, in the case of lytic viruses, the cell dies quickly and falls apart, releasing hundreds of infectious viruses.

336. (D) Kuru was a degenerative human disease that was found in Papua New Guinea. Infected individuals suffered increasing dementia and death. Upon autopsy, the diagnosis was spongiform encephalopathy, which indicates that kuru is probably caused by the same prion agents that cause scrapie in sheep, elk wasting disease, mad cow disease (bovine spongiform encephalopathy), and Creutzfeldt-Jacob in humans.

Chapter 17: Prokaryotes

337. (C) Since temperature affects enzyme activity, all enzymes within a thermophilic cell function best at thermophilic temperatures. Temperatures above or below the optimal temperature for these enzymes will decrease enzyme activity, and the farther away from the optimal temperature, the greater the depressive effect.

338. (C) *Staphylococcus aureus* is a toxigenic pathogen associated with numerous conditions such as toxic shock syndrome and staphylococcal food poisoning. *Streptococcus pyogenes* is similarly pathogenic and causes strep throat and rheumatic heart disease. *Neisseria gonorrhoeae* is a cause of pelvic inflammatory disease and gonorrhea. Only the first two are Gram positive, only the first can be considered normal flora, and none are obligate intracellular parasites. All, however, appear as cocci under the microscope.

339. (B) For decades the five-kingdom classification system was commonly used to classify the most major groups of organisms and was based primarily on morphology and physiology. However, as genetic techniques developed and molecular analysis flourished, the relatedness of DNA sequences became more significant factors. When genetic relatedness (or lack thereof) is considered paramount, the variations between all animal species become relatively minor compared to those seen in all prokaryotes. In fact, genetics indicate that about 35% of the world is archaebacteria, 35% is eubacteria, and 30% includes everything with a nucleus.

340. (E) The organisms that are the most common cause of meningitis in children very often reside somewhat normally in the upper respiratory system. These can pass through the eustachian tubes and precipitate middle ear infections (otitis media). Occasionally they progress onward and cause meningitis. *Neisseria meningitidis*, *Haemophilus influenzae*, and *Streptococcus pneumoniae* are the three primary bacterial causes of this condition.

341. (D) The kidneys and urinary system reside outside the body, topologically speaking. This is because an organism can enter the urethra, pass through the urinary bladder, work its way up through the ureters into the calyx and nephron tubules of the kidney, and never pass through any tissue or breach any epithelial barrier. Many organisms are capable of doing just that, but only *Leptospira* possesses the very unusual axial filament.

342. (D) The mnemonic "King David Came Over For Good Steak" is an easy way to remember the major taxa: kingdom, division (or phylum), class, family, genus, and species. However, there are many other intervening levels that may be added as necessary, such as superfamilies or subspecies. One of these occasionally added for bacterial taxonomy is the tribe (and sometimes subtribe), which falls between order and family.

343. (B) Almost any bacterium, given the opportunity and a failing human immune system, can use humans as a food source. However, the ones that are given the greatest opportunity, have the most to gain, and generally have the most pathogenic mechanisms are the heterotrophs, which are incapable of feeding themselves and require organic materials upon which to feed.

344. (E) All cells have a cell membrane. Many cells are protected by cell walls usually composed of a polysaccharide such as cellulose, chitin, or peptidoglycan. Only eukaryotic cells have organelles such as mitochondria, chloroplasts, and nuclei, which excludes all prokaryotes.

345. (A) DNA in nucleated cells is organized into numerous chromosomes. These chromosomes are linear and tightly wound around spools of proteins called histones. The genomes of bacteria, on the other hand, are much simpler and lack the histones. The bacterial genome consists of circular double-stranded DNA, and many bacteria also have smaller circular nongenomic DNA in the form of self-replicating plasmids.

346. (D) Bacteria are highly successful organisms that have been able to thrive, or at least survive, in conditions that are extremely hostile to most other forms of life. This includes boiling mud pots, salt flats, and even sedimentary rock several kilometers underground. The term is used to describe the "found everywhere" distribution.

347. (B) The vast majority of bacteria encountered by humans on a daily basis are classified as heterotrophs. In fact, the number of bacteria that colonize humans within and outside our bodies exceeds the number of our own human cells by at least 10:1 because we are such a good food source for them. The archaebacteria are an ancient form of bacteria that have survived because they have been able exploit extreme environmental conditions of temperature, radiation, or desiccation.

348. (A) All prokaryotes lack a nucleus. All eukaryotes have a nucleus. Eukaryotes include protozoans, fungi, animals, and plants, as well as most forms of algae. Of the organisms listed, the only bacterium is *Staphylococcus*. While the others are all microscopic, they are all nucleated protozoans, or protists.

349. (E) All except one of the organisms listed are human pathogens. The one exception, *Thermus aquaticus*, is a thermophilic bacterium that is the primary source of the DNA polymerase used in the PCR process. The sexually transmitted disease in this case is syphilis, and the causative agent is the spirochete *Treponema pallidum*.

350. (A) If you consider that all eukaryotes are organized within one of the three superkingdoms, then they are normally subcategorized into plants, animals, fungi, and protozoans. The first two of these are multicellular and more complex than the others. Fungi generally form large fruiting structures as a result of sexual reproduction. Bacteria lack nuclei and thus are not used in the comparison.

351. (D) The bacterial genome (we normally do not use the word *chromosome* for prokaryotes) is a single circular double-stranded DNA molecule. While a rapidly growing bacterial cell usually contains more than just a minimal single copy because it is continually undergoing DNA replication, it is considered haploid. Bacteria also are frequently infested with much smaller but similar self-replicating structures called plasmids, which need not be present for bacterial survival but frequently carry genes that provide some benefit to the host cell.

352. (B) Under the normal replication process replisomes work in opposite directions along the circular bacterial genome, starting from the origin of replication (ORI). With semiconservative replication, the structure seen with electron microscopy about halfway through the process resembles the Greek letter theta (θ) and is thus known as theta mode replication.

353. (E) The presence of a bacterial fertility plasmid causes the formation of a conjugation tube through which a copy of the plasmid can pass from cell to cell, converting the recipient from an F^- (infertile) cell to an F^+ (fertile) cell. Occasionally, the plasmid becomes integrated into the bacterial genome, converting the cell from F^+ to Hfr^+ (high frequency of replication). Conjugation of an F^- cell with an Hfr^+ cell allows not only the movement of the plasmid but also the transfer of the attached copy of the genome as well, and this transfer occurs at a predictable rate and in a consistent sequence of genes. Interrupted mating experiments used this process to map the *E. coli* genome.

354. (E) There are several natural mechanisms in which DNA can move from bacterial cell to bacterial cell. When a cell is disrupted, commonly at death, the DNA within the cell is released into the surrounding medium. This unattached DNA can be randomly acquired by nearby bacterial cells in a process known as transformation. It is known as transformation because the additional genes frequently change, or transform, the bacterial phenotype.

355. (C) There are numerous mobile genetic elements found within cells that permit the horizontal transfer of genetic material. Among these are insertion sequences (700–1,400 bp in length with inverted repeats at each end), plasmids (independently replicating circular DNA 20–200 kbp in size), and bacterial transposons (2.5–14 kbp in length with an insertion sequence at each end).

356. (C) There are several natural mechanisms in which DNA can be horizontally transferred from cell to cell regardless of the species. One is transformation in which naked uncoated DNA is acquired by a cell from the surrounding medium. Another, transduction, uses bacteriophages to move not only the viral genome but added bacterial DNA as well.

357. (B) Bacterial pili (also known as fimbrae) permit the attachment of one cell to another. Bacterial flagella provide a means for cellular motility. One pathogenic mechanism that enhances the survival of bacteria within a host is the ability to produce siderophores, which scavenge vital iron from host tissues that permits an increase in the bacterial growth rate. Bacterial polysaccharide capsules not only provide protection from phagocytosis but also provide a food storage mechanism. Cilia are structures produced only by eukaryotes.

Chapter 18: Protozoans

358. (C) Protozoans, also known as protistans or protists, are nucleated, single-celled, motile, and generally nonphotosynthetic organisms. Most commonly associated with pond amoebas, paramecia, and ciliates, the group also includes euglenoids, dinoflagellates, and diatoms. Bacteria, because they lack a nucleus, are not included.

359. (D) Malaria is a disease caused by a sporozoan parasite that is spread through the bite of a female mosquito. The initial form released from the mosquito salivary gland first takes up residence in the host liver and then later begins entering host erythrocytes and feeding on the hemoglobin. Later, when proper host proteins are present, the parasite gametes swarm to the upper layers of the skin where they are ingested by another mosquito. The gametes mate, form a cyst in the mosquito stomach, and another form later migrates to the salivary gland to repeat the life cycle. There are several species within the genus *Plasmodium*.

360. (A) A parasitoid is a parasitic wasp that lays its eggs on another insect. When the wasp eggs hatch, they feed on the unsuspecting host, eventually killing it and then using its body as a shelter as they pupate before releasing adult wasps. All of the other organisms are unicellular protozoans.

361. (B) Three of the listed organisms are bacteria. One of the remaining, *Plasmodium* sp., includes the sporozoan parasite species that cause malaria in humans and related disorders in other animals and are spread through the bite of mosquitoes. The other protozoan, *Entamoeba histolytica*, is an amoeboid human pathogen that is the causative agent of the potentially lethal amoebic dysentery.

362. (D) There are two forms of protozoan slime molds. The first, the acellular slime mold, forms a large, usually orange or yellow, coenocytic cell mass that spreads through moist organic debris such as leaf litter and can be cultured easily at home. The second, the cellular slime mold, lives in soils primarily as an amoeba. Under certain conditions, the independent cells will congregate into a generally small but visible slug stage. From this base a fruiting body will grow upward so that the spores formed at the apex can be disbursed in the wind.

363. (A) The protozoan genus *Trypanosoma* is the causative agent for the disease generally known as trypanosomiasis. In the Old World it is better identified as African sleeping sickness. The vector that spreads this obligate parasite from host to host is the dreaded tsetse fly.

364. (A) An abrasive is a material that can be used to abrade, or scrub away, materials from a surface much as sand is commonly used to abrade a surface of rust prior to painting. When added to toothpaste, very small particles act to help remove surface plaque. The silica-based cell walls of diatoms, long dead and packed into sedimentary rock, are commonly used for this purpose. This material is also used to filter pools in the form of diatomaceous earth.

365. (E) Protozoans commonly have at least two life stages: the vegetative form during which the cell metabolizes, grows, and divides, and the dormant survival stage that the cell will assume when the environment becomes too hostile. The former is known as the troph, or trophozoite, stage and the latter the cyst stage.

366. (E) Amoebas may appear to be rather helpless because they are only slowly motile and have no protection other than their tough cell membrane called a pellicle. Some amoebas are capable of producing frequently grandiose-shaped silica-based protective structures called tests. The amoeba can extrude some of its cell structure through openings in this test in order to feed.

367. (C) Under highly favorable conditions, many forms of algae will reproduce exceptionally rapidly, producing what is known as an algal bloom. As conditions change, the dead and dying organisms tend to sink to the bottom of the waters in which they live, and eventually the buildup of these cells forms large layers of sedimentary rock. The white cliffs of Dover were initially formed in this way from the innumerable bodies of dead diatoms.

368. (B) The word *pseudopodia* comes from the Greek roots *pseud-*, meaning "false," and *pod-*, meaning "foot," thus "false foot." Amoebas lack the cilia and flagella of other protozoans that provide them with efficient means of locomotion. An amoeba can move by using its cytoskeleton to contract the cell surfaces on the trailing end, forcing the cytosol and surrounding pellicle to expand forward in the direction of desired travel, and this leading structure is called a pseudopodium.

369. (D) Chlorine is a topical disinfectant effective against a wide variety of infectious and toxic agents, but it cannot be used internally because it would have the same damaging effect on the intestinal epithelium. Penicillin is an antibiotic used against bacterial infections, and miconazole is an antimycotic used to counter fungal infections. Aspirin has no effect on microorganisms other than to relieve the pain or fever caused by their presence. Flagyl is a commonly prescribed antiparasitic agent.

370. (A) Gametes are haploid cells used for sexual reproduction in fungi, plants, and animals. Mycelia are masses of filamentous fungal cells. Lysosomes are digestive structures found in protozoans and animals, and nuclei are found in all of the choices. Only kinetoplasts, specialized cellular structures associated with cell motility, are found in certain protozoans.

371. (E) Malaria is a blood-associated protozoan disease caused by *Plasmodium* sp. Giardiasis is an intestinal disease caused by protozoans spread through contaminated water. Elephantiasis is caused by microfilarial worms of the genera *Wuchereria* and *Brugia* and is so named because these parasites block lymph ducts and cause a huge edematous expansion of the lower extremities. Myiasis is actually a condition in which fly maggots live in a living host, a process that can be used for dead tissue debridement. Leishmaniasis is a protozoan disease spread by fly bites.

372. (B) Most protozoans that infest the digestive tract of animals can exist in two different forms: the vegetative form known as a troph and the survival form known as a cyst. Clinical diagnosis is usually based upon the microscopic examination of fecal matter for the presence of either of these forms. Species identification is based upon size, internal structure, and overall morphology.

373. (D) Double fertilization is a process unique to plant angiosperms. Binary fission is unique to bacteria, and mitosis is the functional equivalent of cell division for eukaryotes. Meiosis is the cell division mechanism that leads to the production of gametes prior to sexual reproduction. Schizogony is uniquely a protozoan mechanism that involves cellular fragmentation of the parasite within the host cell.

374. (A) While protozoans may all appear dangerous to human or animal health, that is because we tend to find only those that are making us sick. However, the vast majority of protozoans are not parasitic but rather are widely dispersed throughout the environment in free-living form. It should be pointed out, however, that many of these free-living organisms will happily begin using humans as a food source if given the opportunity.

375. (D) While some protozoans, notably euglenoids, diatoms, and dinoflagellates, can be photosynthetic, the majority have metabolism much as humans have in that we are heterotrophic. This means that these cells require the intake of preformed organic compounds as their source of carbon, which also means that some other cell has to die in order for these cells to live.

376. (E) Protozoans, as eukaryotes, are capable of asexual reproduction in the form of mitosis. Another form of asexual reproduction, merogony (also known as schizogony) occurs when certain protozoans undergo multiple cellular divisions that are not mitotic within a host cell. Many, including most of the ciliates, are also capable of a somewhat sexual form called conjugation, in which conjugated cells exchange micronuclei. Protozoans are not capable of sexual reproduction as a result of meiosis, although some form gametes through other mechanisms.

377. (C) In the New World trypanosomiasis is spread by reduviid bugs, also called kissing bugs because of their preferred nighttime biting locations around the mouth. After their blood meal, they rotate their bodies 180 degrees and defecate on the bite site, which then allows the parasite to penetrate and infect a new host. This disease is most commonly seen in the tropics and is commonly identified as Chagas disease.

378. (C) One of the more primitive protozoan groups is thought to be so because it lacks Golgi bodies and mitochondria, a characteristic that allows it to survive under anaerobic conditions. These organisms form a barrier covering the epithelial surface of the intestinal lumen in animals, preventing the absorption of lipids, which causes intestinal distress and diarrhea. The genus of these pathogens is *Giardia*, and they are the cause of giardiasis.

Chapter 19: Fungi

379. (B) There are four divisions within the kingdom Fungi. These divisions are based on the sexual reproductive fruiting structures. Mushrooms, which fall within the division Basidiomycota, are so classified because spores that are formed following sexual recombination are borne on a short stalk known as a basidium, which is found suspended from the gill-like structures on the undersurface of the mushroom cap.

380. (C) Fungi have a unique digestive mechanism whereby they secrete their digestive enzymes into the surrounding medium. This permits the extracellular breakdown of organic compounds into the vegetative cells. This can be an essential form of decomposition when the fungi feed upon dead material, in which case they are known as saprobes. However, some fungi are opportunistic and even pathogenic when they feed upon living material, in which case they are parasitic.

381. (A) Some fungi are capable of feeding on the dead, outer, keratinized, epithelial layers of the skin. These organisms are known as dermatophytes and can cause tinea capitis (ringworm), tinea cruris (jock itch), and tinea pedis (athlete's foot).

382. (E) Fungi are found in numerous food products. They are responsible for the fermentation of alcoholic beverages such as beer and wine, and the fermentation of foods such cheeses, sausages, and bread (*Saccharomyces cerevisiae*). Additionally they can be found in grocery stores in the form of the common table mushroom *Agaricus bisporus*.

383. (B) Organisms are not capable of such transitions as from bacteria to protozoans or from protozoans to yeasts. Organisms commonly have two copies of every gene as implied by the term *diploid*. For fungi, however, the term *dimorphic* (literally "two forms") means that most species can be found in the form of single-celled yeasts or multicellular filaments, and they can switch from one to the other depending upon the environmental conditions.

384. (D) Fungi are officially classified within the major divisions based on the structure that bears the sexually formed spore. Some species, primarily those that are pathogenic, have not yet had this structure identified. These "none of the above" organisms are classified together in the catchall division Fungi imperfecti (*imperfect* meaning "without known perfect, or sexual, stage") or Deuteromycota, which contain all the deuteromycetes.

385. (E) Fungi do not have sexes; they have mating types, with some fungi known to have as many as eight of these types. Fungal sexual reproduction can take place when one member of a species with a certain mating type makes contact with a member of the same species but of a different mating type. Sexual recombination takes place, and a structure is produced that holds the resultant spores. If this structure is a larger sporelike structure with a thickened wall, it is known as a zygospore and the species belongs within the Zygomycota.

386. (A) Fungal spores are frequently seen within saclike structures, and their appearance can be confusing for classification purposes. Some species will form the sacs as an asexually produced structure. Commonly these structures are called sporangia, and they contain sporangiospores. However, these sacs are filled with spores produced by sexual recombination, in which case the sac is known as an ascus and the organism is classified within the division Ascomycota. One key differentiation between an ascus and a sporangium is that an ascus will contain only an even number of ascospores, usually eight or less, while a sporangium will usually contain many more.

387. (E) Fungal spores formed asexually are identified exclusively based on their morphology. While none of these will ever be classified under the sexual spore form names of ascospore, zygospore, or basidiospore, they may be classified with numerous descriptors such as arthrospores, chlamydospores, conidiospores, phialospores, porospores, blastospores, macroconidia, and microconidia.

388. (E) Multicellular algae are classified as plants, and as such, they contain chloroplasts and have a cell wall usually composed of cellulose. Fungi fall within a separate kingdom, lack chloroplasts, and have cell walls composed of chitin. Since both are eukaryotes, both have mitochondria, which therefore cannot be used to distinguish between the two.

389. (D) Beer and bread are foods in which a key productive process involves the fermentation of sugars by yeasts: in the former, to produce ethanol and in the latter, to produce carbon dioxide gas, which causes the leavening. Sauerkraut and cheese also contain fermentation products of fungi, but additional fermentation produced by bacteria is also important.

390. (A) The cell walls of bacteria, algae, plants, and fungi are all composed of polysaccharides. In most bacteria, the basic material is peptidoglycan. In algae it is often cellulose, agarose, or carrageenan. In plants, it is mostly cellulose. In fungi, it is chitin, which also happens to be the basic material that comprises the shells of shrimp and lobsters.

391. (B) Lichens have been observed all over the planet and are commonly the first organism to colonize newly formed rock in the absence of soil. Until the invention of the microscope, they were thought to be plants and over 40,000 species had been identified. It is now known that lichens are actually close associations of fungi, which provide protection, and algae, which provide food, that live in symbiosis.

392. (B) Most fungi are capable of both sexual and asexual reproduction. However, the mechanism of asexual reproduction within yeast is unique. A new nucleus is formed within the mother cell, and a new cell begins to form with a bulge on the cell wall. This new wall continues to expand, filling with cytosol and the appropriate amount of cellular organelles from the mother cell. Eventually the new nucleus also enters the new daughter cell, and a new wall forms and begins to pinch off the new smaller cell. This process is known as budding.

393. (C) A single strand of a filamentous fungus is known as a hypha (singular). When they are clustered with more than one strand, they are called hyphae (plural). When there are enough hyphae to become visible to the naked eye, such as forming a colony on a petri plate, the colony is referred to as a mycelium (singular). When more than one mycelium is present, the group is referred to as mycelia (plural).

394. (A) The earth's ecosystem is dependent on the movement and recycling of materials across the planet. Water evaporates from oceans, rivers, and lakes; moves through the atmosphere; and then returns as rain or snow elsewhere. In the same way, carbon in the air and water is captured in organic matter, is stored as biomass, is broken down, and finally returns to be reused. Fungi, which are major components in the breakdown portion, permit this cycle to continue. Without them we would all be 300 meters deep in fallen timber.

395. (D) While fungi are classically divided into the large divisions of ascomycetes, basidiomycetes, zygomycetes, and deuteromycetes, a fifth group, the oomycetes, are also considered fungi because of their dominant fungal characteristics. They are unique, however, because they resemble animals in that they produce motile zoospores.

396. (B) Each year immigrants, usually recently arrived from Southeast Asia, fall ill and die after collecting and eating wild mushrooms. This is because a commonly eaten variety from their homeland greatly resembles the European and New World's *Amanita phalloides*, or Death Cap, which is highly toxic.

397. (A) Many of the edible varieties of fungi appear above the ground because of sexual recombination taking place underground between different mating types that live in the soil. Truffles are different in that the fruiting structures of these mycorrhizal fungi are produced in the soil at the base of several varieties of trees. These are highly prized by gourmands and are commonly found by their scent detected by "truffle pigs" or dogs.

398. (E) While the basidiomycete structures called mushrooms and toadstools are traditionally associated with fairy rings and other fanciful stories, the fact remains that these aboveground structures are simply permitting the dispersal of spores that are the result of sexual recombination taking place in the soil.

399. (C) The name "Fungi imperfecti" (imperfect fungi, where "perfect" is a sexual form) refers to the fact that the species that fall within this group either lack compatible mating types or that those different types have never been brought into contact while observed by researchers. This means that these fungi can reproduce only asexually, and for fungi that means mitosis.

Chapter 20: Plants

400. (D) Plants are known for their characteristic alternating generations life cycle that consists of both sporophyte and gametophyte forms. Gametophytes are identified as such because they are haploid forms that eventually produce haploid gametes that are produced in gametangia.

401. (B) In the more primitive plants, the gametophyte and sporophyte generations are more clearly visualized when compared to the higher plants where the gametophyte stage is reduced to rather small structures within the sporophyte plant. Bryophytes such as mosses form a distinct spore-bearing structure called a seta.

402. (E) A cotyledon is an embryonic leaf that can be found within a seed. Seed-bearing plants fall into one of two groups: those that produce one cotyledon (monocots) and those that form two (dicots). While the cotyledons of dicots are more likely to be identified as they emerge from the soil following germination, those of monocots stay within the seed.

403. (A) Rhizomes are plant structures that are stems that grow underground and commonly are capable of being separated from the plant of origin and propagating a new plant. Fronds are the leaflike structures that bear the spore-generating sori on the underside of the leaf. Both are best associated with sporophyte structures of ferns.

404. (A) Flowers and stems that are produced by monocots differ in structure from those that are produce by dicots. The stems of monocots are always herbaceous, while those of dicots may be either herbaceous or woody and capable of multiyear growth. The petals of monocot flowers tend to be arrangements of threes or multiples of threes; those of dicots are not.

405. (C) Angiosperms, or flowering plants, are unique in that they rely on a double fertilization process for the formation of their seeds. One fertilization produces the diploid plant embryo. The second simultaneous fertilization of two degenerate polar nuclei (produced through meiosis along with the egg) with a sperm produces the triploid endosperm structure of the seed that provides nutrition to the embryo.

406. (A) In the more primitive plants, such as those that are nonvascular or vascular but seedless, the sporophyte and gametophyte forms are readily identifiable. In the case of vascular seedless plants, the structures of these two generations hardly resemble each other and only close observation uncovers the linkage. Ferns are seedless vascular organisms that produce both diploid fronds and the much smaller haploid prothallus plants.

407. (B) Flowers are the reproductive structures of angiosperms. The stamen produces the sperm cells, packaged as pollen, in the anther on the end of a filament. When released, the pollen is collected on the stigma. This begins to produce a pollen tube that grows through style into the ovary, which contains the ovule that houses the egg. When connected, the sperm passes through the tube to fertilize the egg.

408. (E) Seeds are the reproductive products of the more advanced plants and are not seen in nonvascular and the more primitive vascular plants. Such seedless plants include mosses, liverworts, hornworts, ferns, lycophytes, and horsetails. Seeded plants include the gymnosperms and angiosperms.

409. (E) In the majority of life-forms on the earth, an organism is either diploid or haploid. Diploid organisms may produce haploid cells called gametes, but the function of these cells is solely to fuse with another to produce a diploid zygote, and these gametes are not capable of independent growth. The exception is plants, which have a life cycle that passes through both haploid and diploid forms.

410. (C) A fruit is a structure that carries seeds and is designed to encourage dispersal by animals. These fruit may contain one or more than one seed. While gymnosperms such as conifers, cycads, and ginkos produce seeds, only angiosperms produce fruit.

411. (E) An organism that is autotrophic is capable of producing its own carbon-based organic materials, usually through the energy-gathering mechanism of photosynthesis. These organisms may be unicellular, such as the protistan euglenoids or dinoflagellates, or they may be in the multicellular form of plants.

412. (D) Regardless of what an organism is, it has interconnective relationships with all other organisms within its habitat. Some of these relationships are neutral and some are more intimate. The most intimate interspecies relationship is symbiosis where two organisms are so dependent on each other that they cannot survive without the presence of the other. Such a relationship is seen between fungal and algal species in the form of lichens.

413. (E) Because plants form rigid cell walls, once formed, the cell must remain in place and connected to adjacent cells for life. Because of this, plants have cellular reproduction and growth only in highly localized areas called meristems. These can be found in areas that produce peripheral growth in stems and trunks and at the tips of stems and roots. The apical meristem in roots actually is found behind a protective layer of nongrowing cells called the root cap. As the meristem produces new cells, this cap is forced downward, thus burrowing through the soil.

414. (C) Plants are referred to as vascular if they have specialized cells that permit the transport of fluids great distances, supporting plant growth. The tissues that conduct newly synthesized food material from the leaves to feed the rest of the plant are collectively called phloem. Those that conduct water from the roots to the leaves are called xylem. The conductive cells that make up xylem include tapering and overlapping tracheids and the more strawlike, end-to-end connecting cells called vessel elements.

415. (A) Plants have a problem: they live in a very arid environment (compared to plants that live in water), so they need to hoard fluids. However, they also need access to the atmosphere, so the process of evaporating water molecules into the air can assist in bringing water up from the roots. This process is called transpiration, and this release is permitted through guarded openings in the leaves called stomata.

416. (E) All plants have two energy-related cellular organelles: chloroplasts and mitochondria. Chloroplasts capture the energy of light, use it to drive the cellular process that produces energy-containing organic compounds, and in the process, release oxygen as a waste gas. Mitochondria take these organic compounds and their derivatives and break them down through catabolism to release their stored energy and drive metabolic processes required by the cell. Because the energy produced within the mitochondria requires the flow of high-energy electrons, there must be a final electron acceptor available to serve as an electron sink. This final electron acceptor is oxygen.

417. (C) The longer a woody, plantlike tree lives, the greater the girth of the trunk becomes. This is because the living portion of the tree is actually on the periphery just under the protective layers of the bark, and each year, during conditions that permit growth, this layer grows out farther and farther from the center of the trunk. These areas of lateral growth are known as lateral meristems.

418. (A) Mitochondria are energy-producing organelles found in almost all nucleated cells, including plants. Also found in all nucleated cells are endosomes, which are required for vesicular transport of materials from the endoplasmic reticulum and Golgi body to the cell membrane. Vacuoles, which contain water, and chloroplasts, sites of photosynthesis, are also required by plants for cellular function. Lysosomes, necessary for the degradation of organic materials by heterotrophs, are not needed by plants because they are autotrophic.

419. (A) Plants are known to have some of the greatest survival rates on the planet. Some seeds, which contain living but dormant embryos, have been found to still be capable of germination and growth over 30,000 years after formation. Some groves of trees, all genetically identical and each growing as a portion of the other, date back over 80,000 years. However, the record for the longest surviving individual is a bristlecone pine tree, a gymnosperm, in California that has over 5,000 annual growth rings.

420. (B) Meiosis that occurs within the ovule of angiosperms is rather unusual. Instead of the standard process of producing four haploid cells, each containing an equal amount of genome, cytosol, and organelles, the process results in an uneven distribution of these materials. One megaspore mother cell produces eight nuclei, one of which acquires the bulk of the cytoplasmic contents, two nuclei that acquire the rest and are referred to as polar nuclei, and five nuclei that receive no supporting materials and fully degenerate.

Chapter 21: Animals

421. (E) Animal tissues, as with all multicellular organisms, are recognized as existing within different levels of organization and function. At the lowest and simplest level are the individual cells. Similar cells commonly function together in the form of tissues, and various tissues function together to form small glands or larger organs. Various organs then function together as organ systems, all of which function to maintain the homeostasis of the body.

422. (A) After an egg is fertilized, it becomes a zygote. This zygote then undergoes a period of rapid cellular division and eventually forms a spherical structure with an internal cavity. Some cellular specialization begins to occur, and three basic tissues eventually develop: endoderm, mesoderm, and ectoderm. From the ectoderm, later epithelial tissues develop.

423. (E) In order to understand the functioning of the body, students of anatomy virtually subdivide it into various halves with a convention known as planes of division. The anterior half of the body is separated from the posterior half by the frontal, or coronal, plane. The superior, or upper, half is separated from the inferior, or lower, half by the transverse plane. A parasaggital plane divides right (dexter) from left (sinister), while the more specific term *midsaggital* refers to the division in equal halves.

424. (B) There are estimated to be at least 35 animal phyla. These have traditionally been separated based on phenotypic traits, but more recently genetics and developmental evidence have had their say as well. Most of the major tissue groups are observed even in the simplest animal groups. The simplest, or lowest, animals to possess nervous tissue are the cnidarians, which include hydra and jellyfish.

425. (C) Flukes (trematodes) are a large group of animals that are a subset of flatworms but are all parasitic. Parasitic worms of all forms (flukes, tapeworms, and nematodes) have complex life cycles, mostly involving more than one host species. Typical for trematodes is *Schistosoma mansoni*, whose life cycle includes humans as a definitive host and snails as an intermediate host, and which causes the liver-involved disease schistosomiasis, or bilharzia.

426. (D) Animals, particularly the ones with which most people are familiar, have a central body cavity, or coelom. Mammals, whose coeloms are fully lined with mesoderm-derived tissues, are known as coelomates. Some species have a coelom that is partially lined with mesoderm-derived tissue and are called pseudocoelomates, which includes molluscs. More primitive organisms such as flatworms lack a coelom and are thus acoelomates.

427. (D) Bivalves include organisms that are protected by a hinged shell. These include oysters, clams, and scallops. Brachiopods, while also fitting this basic description, have significant genetic and ancestral differences and are not as common as bivalves. Bivalves are essential in estuaries because their filter-feeding mechanism cleanses the water. However, this mechanism also means that bivalves may serve to collect pathogenic bacteria that can cause food-borne diseases and concentrate neurotoxins produced by some dinoflagellates.

428. (B) Tunicates are marine nonvertebrate chordates that include sea squirts. In their larval form, they resemble amphibian tadpoles. They are called tunicates because they produce a protective covering that reminded researchers of a human tunic. This tunic is made from a material unusual for an animal, the polysaccharide cellulose.

429. (E) Based on fossil evidence, the first animals to colonize land were large short-necked, salamander-like animals with short tails. These labyrinthodonts are thought to have derived from lobe-finned fish that were able to occasionally venture onto land to perhaps feed or escape predators. These animals disappeared and were replaced by reptiles about 300 million years ago during the Carboniferous period.

430. (C) One of the biggest problems organisms of all types had when moving into terrestrial habitats from marine or freshwater environments was the comparatively arid atmosphere. In order to survive by retaining sufficient fluids, epithelial surfaces surrounding the organism had to be able to prevent water loss. In reptiles we see the success of such efforts as a covering of scales.

431. (E) The most commonly identifiable characteristics of mammals are their covering of hair and the production of milk to support the early growth of their young. Lesser known characteristics include the presence of a muscular diaphragm and teeth that are differentiated by form and function. A placenta, while frequently also thought to be possessed by all mammal species, actually is not found in monotremes or marsupials.

432. (A) There are two categories of mammals that lack a placenta. The most primitive of these are the egg-laying monotremes, which includes the platypus and the echidna (spiny anteater). Another group, the marsupials, have live births, but the less-developed young are nurtured by milk and the mother's pouch until they are safely more mature. Marsupials include the kangaroo, koala, wallaby, wombat, bandicoot, and opossum. Most of these can be found in Australia.

433. (A) The fossil record indicates the appearance of small reptiles that possessed longer legs built for running and differentiated teeth. Additionally there is evidence that these were also endothermic and possessed fur or at least hair. These animals, which lived around 200 million years ago during the Triassic period, were thought to be early forms of mammals known as therapsids.

434. (B) Some of the major characteristics that differentiate primates from other animals is their increased brain capacity and their opposable thumb. Additionally they possess generally more acute vision and a greater reliance on stereoscopic vision. It is thought that the cost of this improved vision appears to come at the expense of a loss of olfactory abilities.

435. (B) There is a type of small fish that spawns on the California coast. At high tide the female digs tail first into the sand, and then the male wraps himself around the female and fertilizes the eggs. The eggs incubate until the next high tide, at which time the newly hatched young are washed out to sea. People are allowed to collect the spawning fish on selected dates at night by hand for their own consumption, and these "grunion runs" are very popular.

436. (E) Most organisms, including animals, plants, fungi, and even bacteria, have a daily physiologic cycle that controls their activities and that is known as a circadian rhythm. In animals this rhythm is controlled by hormones secreted by a single gland in response to light perception. This master controller is found at the base of the brain near the pons and is called the pineal gland.

437. (D) After the birth of any mammal a physiologic and psychologic bond is formed between the mother and child. This bond, a form of imprinting, helps insure the mother's immediate support of the child to enhance survival through early infancy. While the mechanism that triggers this imprinting varies from species to species, in sheep it has been shown to be triggered through the sense of smell.

438. (C) Animals learn through a variety of mechanisms. Most learn by experience, but this can be costly. Many others can also learn by observation, as was seen decades ago when chickadees passed on to each other their success at gaining access to high-energy cream when they pecked their way through the paper caps on home-delivered milk bottles. Fewer still can learn by thought processes alone, also known as insight learning. Primates, in general, and humans, in particular, are the most adept at this.

439. (A) Terrestrial animals tend to have daily biologic cycles based upon the perception of light, whether they be nocturnal or diurnal. Subterranean animals, which obviously lack these visual clues, do not. Parasitic animals tend to follow the rhythms of the host, as they have the ability to sense host hormonal and other chemical signals. Marine animals are different, however, because they tend to follow tidal, thus lunar, signals.

440. (C) Nematocysts, also known as cnidocysts, are small subcellular structures that are produced by cnidocytes. These nematocytes are small harpoon-like structures that can be explosively ejected from the host cell to deliver a toxin for the purpose of either capturing prey or for protection. These structures are what cause the sting from jellyfish even when stranded on shore.

Chapter 22: Ecological Principles

441. (D) No organism lives independent of any other organism. Humans are, in fact, one large ecosystem with coinhabitants inside and outside. If counting the respective numbers, the number of cells that contain our own DNA and comprise us as distinctive individuals, and then comparing that number to the combined number of bacterial, protozoan, fungal, and other animal cells that we carry with us, they outnumber us 10:1. We study similar interactions of organisms within an environment when we study ecology.

442. (A) We sometimes have to study smaller groups and their interactions before we can understand the whole. One of these smaller subsets is a community, which considers all of the members of all of the species present within a single habitat. All of the organisms living in and on us on a daily basis would be one such community to consider.

443. (C) A habitat is an environmental area in which an organism dwells. If we were to study all of the species within that habitat and not just that one, then we would be studying a community. If we were to focus on the organisms of just a single species within a habitat, then we would be studying a population.

444. (B) Often one cannot understand the behaviors or relationships between organisms without also considering the effects of the environment. Doing so leads to the study of an ecosystem. An example would be when the interactions of organisms and the numbers within a population under drought conditions are different than those observed during periods of normal water availability.

445. (D) Organisms have specific nutritional requirements that must be met in order to survive within a community, and this means not just the animals and plants that might be easy to observe. Based on their feeding behaviors we classify animals as herbivores (those that exclusively eats plants), carnivores (those that exclusively eat other animals), and omnivores (those that eat both plants and animals). A finch is an example of the first, the cat and osprey are examples of the second, and the bear, which will happily eat either fish or small game animals or roots and berries, is an example of the third.

446. (B) When conditions are very unfavorable, the number of species within a community will generally decline. The opposite occurs when conditions become more favorable. This means in areas where there is a large amount of energy available for the formation of biomass, the numbers of species, or biodiversity, increases. The biome that receives the greatest amount of energy and water resources lies within the tropics in the form of the rain forest.

447. (C) The foundation of any ecosystem is the number of producers (plants) that are available to convert the energy from the sun into chemical energy in the form of biomass. The greater the number of consumers relative to producers, the less energy is available because of metabolic waste. The larger the ecosystem, the greater the diversity, the greater the resilience to stresses, and the greater overall stability.

448. (D) Some relationships between species are neutral, and they have very little effect on each other. Commensals, however, benefit from another species while having a neutral effect on it. Some species compete with each other, and the gain of one means loss to the other. Parasites and predators take full advantage of another species to that other species' detriment. At the other extreme lies symbiosis, where the loss of one causes the loss of the other, and the gain of one is also the gain of the other.

449. (A) Predators and parasites are greatly alike. Both represent relationships where one organism gains to the detriment of another. The difference between the two is a matter of location: a parasite lives in or on its prey (such as a tapeworm or louse), while a predator lives apart (such as a lion or mosquito).

450. (E) Flies can live apart from humans just as humans can live apart from flies. Although flies carry organisms that can act as human parasites, they are not parasites themselves. Even myiasis, a condition where fly larvae live within dead tissue such as a wound (and still have some current clinical uses), the condition is not to the human detriment. Therefore, the relationship between humans and flies is purely a commensal one.

451. (E) When there are limited resources, individuals must compete with each other to acquire those resources. If the competitors are from different species, one generally is able to acquire a greater share than the other because of differences in the ability, or need, to exploit it. This relationship is known specifically as interspecies competition.

452. (D) Because the needs of an organism are dependent on the needs of cellular metabolism to support the homeostasis of the organism, each species has a unique combination of conditions that make a certain community an ideal place to thrive. The ecosystem that provides the best or ideal support for a certain species is known as its niche.

453. (B) A parasite is an organism that feeds on a host while also living in (endoparasite) or on (ectoparasite) that same host. Because of their unique host requirement, without which a species cannot propagate, some insects fall into the category of parasitoids. These insects deposit their eggs upon other insects. The hatched larvae feed upon or consume the host, after which the larvae pupate and then emerge as adults to renew the life cycle.

454. (C) Some parasites are classified as such, not because of what nutrients they take from a host physically, but what energy they take by way of behavior. Most birds raise their own young from hatchling to fledgling, sometimes even with cooperation from other nonmated birds. The cowbird, however, lays its eggs alongside those in the nest of other birds. The unsuspecting parents then raise the adopted chicks alongside their own, freeing the cowbird to spend its energy in feeding and not raising a family.

455. (A) Predation and parasitism both have energy costs laid upon the host that limit its own growth. Emigration, when some members of a population depart for greener pastures, also costs the population those departed numbers. Resource portioning, when two species compete for a resource, also means that both lose a little of the resource and growth that would have resulted if they had been able to exploit it all. Mutualism, on the other hand, means that there are no losses to either interacting species, because both benefit from the relationship.

456. (D) All organisms within their native habitat have another species that uses them as a food source, as even those at the top of the food chain can be parasitized. This systemic interdependence and biodiversity produces the greatest stability within a community. However, when one species from outside the community is introduced without its natural predator, then that invasive species will tend to outcompete all native forms due to the lack of a natural control on its growth.

457. (C) When bare rock is laid down following a volcanic eruption, there are no organic materials to support life. However, after a short time very slow-growing, autotrophic organisms such as lichens colonize the area. Following them and in a fairly predictable order will come a progression of more sophisticated and diverse organisms that build a constantly changing community upon the resources left by the previous generations. This series of progressive changes is known as community succession.

458. (B) Community succession is a series of progressively more complex and more interdependent populations within a community. The first community to appear, the primary succession, provides changes that support the following groups of species, known as the secondary succession. The number and types of organisms within the community will progressively change until it fully matures and stabilizes as the climax community.

459. (A) While total symbiosis, where each species present is equally dependent on and provides equal benefit to every other species, may appear as an ideal utopia, it ignores the needs of all heterotrophs: in order for a heterotroph to survive, something else must die. Therefore, the probability of such total symbiosis is zero.

460. (E) Success within an ecosystem implies stability within the various populations in terms of numbers and their effects on other populations. While there may be some fluctuations produced by variations within the environment, a climax community is optimally stabilized by zero growth within all populations. If a new species enters and the community stabilizes again, then the new species has become fully adapted to the new community.

Chapter 23: Population Ecology

461. (D) Individuals within a population will be physically distributed throughout a habitat based on a variety of factors, including behavioral preferences, resource distribution, and distribution and density of predators and prey. Thus, some species will congregate around resources or for social interactions in a clumped pattern, some will cover broad ranges in unpredictable random patterns, and some will space themselves fairly evenly in a uniform pattern.

462. (B) The population of a species within a given area is never constant, because the environment and resources are never constant. Thus populations will increase when births exceed deaths or when favorable conditions attract new immigrants from surrounding areas.

463. (A) There are three types of survivorship curves based on actuarial data. The Type I survivorship curve is seen when a species has few or individual offspring with low infant mortality. These offspring are given large amounts of nurturing for a rather longer length of time and tend to have longer life spans. This pattern is typical for large terrestrial animals, including humans.

464. (A) It is estimated that the number of humans on earth will reach eight billion around the year 2025. However, the population took many millennia to reach the first billion sometime in the early nineteenth century. The growth rate was much slower then than it is now, and it took almost 100 years to reach the second billion mark in approximately 1930.

465. (B) If we plot the numbers of a population relative to time, we see the effects of the environment on that population. When the curve is nearly flat with little growth or loss, it represents zero growth. Sometimes the curve initially reflects zero growth, shifts to exponential, and then becomes stabilized at a higher number. This S-shaped curve represents logistic growth.

466. (C) Population numbers depend on a variety of external factors as well as behavioral ones. Some of these factors are not dependent on the density within a fixed area. Such density-independent factors include major disasters such as wars or floods that can cover broad areas equally and without regard to distribution patterns.

467. (A) The growth rate of a population within a fixed area will increase or decrease depending on various environmental and behavioral conditions. Populations will decrease whenever the number of deaths for that species exceeds the number of births or because individuals leave the area by emigrating elsewhere.

468. (C) A Type III survivorship curve includes species that produce large numbers of offspring that receive little or no parental nurturing, experience high infant mortality levels, and generally experience rather short life spans. This is typical for small aquatic animals such as sponges, which release thousands of eggs for random fertilization to the tides in the hope that some will survive to maturity.

469. (D) The population of humans on earth has been increasing exponentially, following a J-shaped curve of population growth, since the mid-nineteenth century. At that time humans numbered around one billion. Two billion was reached by the third decade of the twentieth century, doubling again by around 1974, and most recently reaching seven billion in 2011.

470. (A) Researchers can plot the numbers of a population versus time. If the number starts low and begins increasing slowly, but then the growth rate shoots way up, then the observed J-shaped curve indicates exponential growth, much as we have seen for the human population over the past 150 years.

471. (C) From 1999 to 2011, the human population on earth increased by about one billion people. During that 12 years there were 4,380 days and a total of 105,120 hours. Simple division shows that the population during that time increased at a rate of about 10,000 people per hour. This number does not represent just the number of births, but also accounts for losses due to deaths.

472. (B) A Type II curve, typical for small terrestrial animals, includes species that produce moderate litters of offspring in which the parents provide some early nurturing and for which the mortality rate is fairly constant through their lifetime.

473. (C) Environmental factors can largely affect population size. Some of these factors, such as how effectively predators can acquire prey or a disease or parasitic infestation that might spread from individual to individual, are dependent on the density of the population (think fox in a henhouse or masses packed into a walled city during the plague).

474. (D) When resources are limited, there is competition between individuals for that resource. If the competitors are from the same species, it is known as intraspecies competition. This form of competition is the most intense because the requirements and abilities to exploit a resource are identical.

475. (C) The greater the number of factors contributing to growth that are present, the greater the overall growth rate. Factors that fall into this category include earlier sexual maturity, increased resources that reduce competition, increased desirability of an environment that increases immigration, and the ability of the same population to exploit new habitats within the same area.

182 › Answers

476. (A) When determining the survival rate for a population, it must be understood that life expectancy varies according to the age of the individual. In order to identify these general differences, actuaries cluster individuals of the same age into groups known as cohorts so the data can be better quantitated and evaluated.

477. (E) All life ends in death. The mechanism that produces that death varies from place to place and time to time. While infectious diseases and warfare have trended to reduce populations in large numbers, and while the number of deaths due to cancer has made news, the greatest cause of human deaths remains starvation and malnutrition.

478. (A) The relationship between the environment and a population within a community depends greatly upon the population density. This is because the higher the density, the more focused the draw upon limited resources. Conversely, the more concentrated any highly desirable resources are the more clustered the population will be around those resources.

479. (B) A population will experience increases due to a variety of factors, including increasing birth rates and immigration. But both have to be considered along with the death and emigration rates, and biomass and population densities provide measures only at a point in time. The best measure of population growth is the amount of time a population takes to double in number, and this also permits comparisons to other populations as well.

480. (C) The carrying capacity is the maximum population a given area will support indefinitely, meaning that the population shows zero growth. If the population increases, one or more limited resources will not be able to support the increase and the population will exponentially decline until the carrying capacity is no longer exceeded.

Chapter 24: Communities and Ecosystems

481. (C) Ecosystems comprise the organisms that manipulate the energy flow available to them. This includes producers that initially harvest light or other inorganic energy, which then pass that energy in the form of biomass to consumers, which then pass it on to the decomposers, which recycle the reduced biomass.

482. (E) Carbon, upon which all life depends, is stored in two large reservoirs on earth. It is available in the atmosphere in the form of CO_2 and in the ocean, sediments, and rocks as carbonates. Carbon, bound by organisms through photosynthesis in the form of biomass, is returned to these reservoirs by combustion and respiration. Additionally, volcanic eruptions release additional CO_2 apart from the biomass.

483. (A) While carbon is found throughout the biosphere in various forms, including organic compounds and carbonates, only in the form of CO_2 can it be used for photosynthesis. This is true even in oceans where the dissolved gas is critical.

484. (D) Nitrogen flows through the environment in what has been identified as the nitrogen cycle. This includes the capture of nitrogen gas to ammonia through nitrogen fixation, its inclusion in nitrogenous organic compounds through nitrogen assimilation, its conversion to nitrates and nitrites through nitrification, and their eventual return to the atmosphere through denitrification.

485. (B) As plants draw essential nutrients from the soil, one of the first to be depleted is available nitrogen. This is because although the atmosphere is 78% nitrogen, it is not available to organisms because it must first be fixed into organic form. Nitrogen-fixing bacteria such as *Rhizobium* colonize the roots of legumes such as clover and form root nodules, thus enriching the soil and permitting plant growth because of their activity.

486. (E) Organic compounds are composed of a relatively small number of available elements. Those elements that are predominant in life functions include the gases O_2, N_2, H_2, and N_2. Besides carbon, phosphorus is among the most abundant in cells in the form of nucleic acids and electron-carrying molecules such as ATP. Because it is not available as a gas, this phosphorus is most commonly found as phosphates in ocean sediments.

487. (A) Nitrogen, once fixed into organic form, is acquired by organisms and used in the construction of amino acids, proteins, and nucleic acids. Upon the death of an organism, microbial processes return these nitrogen products back into available ammonia through the processes of decomposition and ammonification.

488. (B) Energy flows through an ecosystem through what are known as trophic (or feeding) levels. The energy of the sun is captured by photosynthetic organisms that are within the first trophic level. Organisms within the second trophic level feed upon these plants, acquiring their energy, and are thus known as herbivores.

489. (C) Ecosystems are continually supplied with energy in the form of light from the sun. This abundant light is captured and converted into biomass through photosynthesis. Because of metabolic losses when this biomass is acquired by organisms at the second trophic level, it is reduced to only 15% of the original, which is further reduced as it continues through the next levels. This means that the greatest biomass resides within the first trophic level.

490. (D) As energy passes from trophic level to trophic level, it is reduced almost by orders of magnitude. Researchers have shown that only 15% of the biomass passes from the first to the second level, only 10% of that is passed to the third level, and only 7% of that continues on to the fourth level. If placed into a graphic form, this data is seen as a triangle or pyramid.

491. (B) Energy available to an ecosystem comes first from the sun. That energy must be converted from electromagnetic form to chemical form by the photosynthesis conducted by plants. From that point on, the energy flow is just a matter of organic acquisition. The most important step in the passage of energy, however, lies in the energy brought into the system in the first place; that is, if the plants aren't there to make it, nobody gets it.

492. (A) Elements and compounds essential for life functions can be found in vast warehouses on, in, and around the planet. Water is found in vast quantities in the oceans, carbon is found in CO_2, and carbonates are in the atmosphere and rocks. The storehouse for nitrogen, required for the formation of proteins and nucleic acids, is in the atmosphere in the form of N_2 gas.

493. (D) Once the energy of the sun is captured by producers and bound up in organic form, it is then passed from trophic level one sequentially on through the consumers at levels two through four. However, through metabolic losses, the amount of energy in the form of biomass is reduced approximately tenfold as it passes from level to level. In practice this means that if humans feed upon grains and a set amount can support a population of 100, and if that same amount of grains is fed to cattle (thereby incurring a loss) and then the cattle are fed to humans, then the number of humans ultimately fed drops to only about 10.

494. (D) The flow of energy through an ecosystem has commonly been referred to as a food chain. However, since omnivores exist and can feed upon producers (placing them at trophic level two), or feed upon consumers that feed upon producers (placing them at trophic level three), or feed upon consumers that feed upon consumers that feed upon producers (placing them at feeding level four), the concept of energy flow is best understood as a food web. Because of this, four trophic levels are the highest number generally required to describe energy flow.

495. (B) The earth has great reservoirs for elements essential for life. When it comes to carbon, we find vast quantities stored in organic form as measured in biomass. However, the bulk of planetary carbon, as found in oceans, is in the form of carbonates, most commonly as calcium carbonate.

496. (E) Water is available all over the planet, with the vast bulk of it stored within the oceans. Evaporation moves significant amounts to other areas where it is deposited and stored in the form of ice, aquifers, and freshwater in lakes, streams, and rivers. Compared to these caches, relatively little is found as easily accessible groundwater.

497. (A) Without the sun, as is seen around deep sea thermal vents, life would be much less common. In those places the primary producers are chemolithotrophic bacteria that capture the energy bound in inorganic compounds and use it as the foundation for their entire ecosystem. When the underlying rocks shift and a vent closes, everything dies. On the surface of the earth, the same function of primary producers falls upon photosynthetic plants.

498. (D) The presence of CO_2 in the atmosphere is essential, as it must be available to plants as the most essential building block of photosynthesis. This carbon, bound by plants in organic form, must be recycled back into the atmosphere in order for life to exist. While the level has varied widely through the ages, the current level of CO_2 in the atmosphere hovers at about 350 parts per million, or about 0.035% of the total atmospheric gases.